Analysis of Sub-synchronous Resonance (SSR) in Doubly-fed Induction Generator (DFIG)-Based Wind Farms

Synthesis Lectures on Power Electronics

Editor
Jerry Hudgins, *University of Nebraska, Lincoln*

Synthesis Lectures on Power Electronics will publish 50- to 100-page publications on topics related to power electronics, ancillary components, packaging and integration, electric machines and their drive systems, as well as related subjects such as EMI and power quality. Each lecture develops a particular topic with the requisite introductory material and progresses to more advanced subject matter such that a comprehensive body of knowledge is encompassed. Simulation and modeling techniques and examples are included where applicable. The authors selected to write the lectures are leading experts on each subject who have extensive backgrounds in the theory, design, and implementation of power electronics, and electric machines and drives.

The series is designed to meet the demands of modern engineers, technologists, and engineering managers who face the increased electrification and proliferation of power processing systems into all aspects of electrical engineering applications and must learn to design, incorporate, or maintain these systems.

Analysis of Sub-synchronous Resonance (SSR) in Doubly-fed Induction Generator (DFIG)-Based Wind Farms
Hossein Ali Mohammadpour and Enrico Santi
2015

Digital Control in Power Electronics, 2nd Edition
Simone Buso and Paolo Mattavelli
2015

Transient Electro-Thermal Modeling of Bipolar Power Semiconductor Devices
Tanya Kirilova Gachovska, Bin Du, Jerry L. Hudgins, and Enrico Santi
2013

Modeling Bipolar Power Semiconductor Devices
Tanya K. Gachovska, Jerry L. Hudgins, Enrico Santi, Angus Bryant, and Patrick R. Palmer
2013

Signal Processing for Solar Array Monitoring, Fault Detection, and Optimization
Mahesh Banavar, Henry Braun, Santoshi Tejasri Buddha, Venkatachalam Krishnan, Andreas Spanias, Shinichi Takada, Toru Takehara, Cihan Tepedelenlioglu, and Ted Yeider
2012

The Smart Grid: Adapting the Power System to New Challenges
Math H.J. Bollen
2011

Digital Control in Power Electronics
Simone Buso and Paolo Mattavelli
2006

Power Electronics for Modern Wind Turbines
Frede Blaabjerg and Zhe Chen
2006

Analysis of Sub-synchronous Resonance (SSR) in Doubly-fed Induction
Generator (DFIG)-Based Wind Farms

Hossein Ali Mohammadpour and Enrico Santi

ISBN: 978-3-031-01373-7   paperback
ISBN: 978-3-031-02501-3   ebook

DOI 10.1007/978-3-031-02501-3

A Publication in the Springer series
SYNTHESIS LECTURES ON POWER ELECTRONICS

Lecture #9
Series Editor: Jerry Hudgins, *University of Nebraska, Lincoln*
Series ISSN
Print 1931-9525   Electronic 1931-9533

Analysis of Sub-synchronous Resonance (SSR) in Doubly-fed Induction Generator (DFIG)-Based Wind Farms

Hossein Ali Mohammadpour and Enrico Santi
University of South Carolina

SYNTHESIS LECTURES ON POWER ELECTRONICS #9

ABSTRACT

Wind power penetration is rapidly increasing in today's energy generation industry. In particular, the doubly-fed induction generator (DFIG) has become a very popular option in wind farms, due to its cost advantage compared with fully rated converter-based systems. Wind farms are frequently located in remote areas, far from the bulk of electric power users, and require long transmission lines to connect to the grid. Series capacitive compensation of DFIG-based wind farm is an economical way to increase the power transfer capability of the transmission line connecting the wind farm to the grid. For example, a study performed by ABB reveals that increasing the power transfer capability of an existing transmission line from 1300 MW to 2000 MW using series compensation is 90% less expensive than building a new transmission line. However, a factor hindering the extensive use of series capacitive compensation is the potential risk of subsynchronous resonance (SSR). The SSR is a condition where the wind farm exchanges energy with the electric network, to which it is connected, at one or more natural frequencies of the electric or mechanical part of the combined system, comprising the wind farm and the network, and the frequency of the exchanged energy is below the fundamental frequency of the system. This oscillatory phenomenon may cause severe damage in the wind farm, if not prevented.

Therefore, this book studies the SSR phenomenon in a capacitive series compensated wind farm. A DFIG-based wind farm, which is connected to a series compensated transmission line, is considered as a case study. The book consists of two main parts:

Small-signal modeling of DFIG for SSR analysis: This part presents a step-by-step tutorial on modal analysis of a DFIG-based series compensated wind farm using Matlab/Simulink. The model of the system includes wind turbine aerodynamics, a 6th order induction generator, a 2nd order two-mass shaft system, a 4th order series compensated transmission line, a 4th order rotor-side converter (RSC) controller and a 4th order grid-side converter (GSC) controller, and a 1st order DC-link model. The relevant modes are identified using participation factor analysis.

Definition of the SSR in DFIG-based wind farms: This part mainly focuses on the identification and definition of the main types of SSR that occur in DFIG wind farms, namely: (1) induction generator effect (SSIGE), (2) torsional interactions (SSTI), and (3) control interactions (SSCI).

KEYWORDS

doubly-fed induction generator (DFIG), subsynchronous resonance (SSR), subsynchronous control interactions (SSCI), wind farm, wind turbine

Contents

CHAPTER 1

Introduction

Nowadays, it is well-understood that the burning of fossil fuels in electric power stations has a significant influence on global climate due to greenhouse gases and causes significant air pollution that negatively affects human health. Additionally, there are widespread concerns about sustainability, given the fact that fossil fuels are a finite resource. For these reasons, in many countries the use of cost-effective and reliable low-carbon electricity energy resources is becoming an important energy policy. Among different kinds of clean energy resources—such as solar power, hydro-power, ocean wave power and so on, wind power is the fastest-growing form of renewable energy at the present time [1] - [5].

The adjustable speed generator wind turbine (ASGWT) has key advantages over the fixed-speed generator wind turbine (FSGWT) in terms of increased power production, reduced mechanical stress, improved power quality, high system efficiency, and reduced acoustic noise. One important class of ASGWT is the doubly-fed induction generator (DFIG), which has become popular with the electric power industry due to its cost advantage over the other class of ASGWT, i.e., fully rated converter-based wind turbines.

Because of increased integration of DFIG-based wind farms into electric power grids, it is necessary to transmit the generated power from wind farms to the existing grid via transmission networks without congestion. However, wind farms, which are frequently located in remote areas, far from the bulk of electric power users, require long transmission lines to connect to the grid. The transmission system options to transmit the wind power over long distances are high-voltage AC (HVAC) [5] or high-voltage DC (HVDC) [6] - [9]. The comparison of these two options has already been studied in the literature [10]. The HVAC solution is viable for distances up to 250 km, and with series capacitive compensation, it may be viable for distances longer than 250 km [5]. Series compensation of transmission lines connecting wind farms to the power grid increases transmittable power over a given transmission line [3], [4].

In the deregulated power market, it is desirable to increase the power transfer capability of existing transmission lines at the lowest cost [11]. Series compensation is considered to be a more economical solution to increase the power transfer capability of an existing transmission line compared to construction of new transmission lines [12] - [14]. Studies show that, in order to increase the transmittable power of an existing transmission line, the total cost of adding capacitive series compensation to a transmission line is much less than the cost of building a new transmission line. As an example, a study performed by ABB reveals that increasing the power transfer capability of an existing transmission line from 1300 MW to 2000 MW using series compensation is 90% less than the cost of building a new transmission line [11].

However, a factor hindering the extensive use of series capacitive compensation is the potential risk of sub- synchronous resonance (SSR) [14] - [23], which may cause severe damage in the wind farm, if not prevented. The SSR in wind turbine generator systems is a condition where the wind farm exchanges energy with the electric network, to which it is connected, at one or more natural frequencies of the electric and mechanical part of the combined system, comprising the wind farm and the network. The frequency of the exchanged energy is below the fundamental frequency of the system. Three different types of SSR in DFIG wind farms have been identified in the literature [24] - [30]:

- Induction Generator Effect (SSIGE)

- Torsional Interactions (SSTI)

- Control Interactions (SSCI)

In the case of SSIGE, the magnitude of the equivalent rotor resistance at the subsynchronous frequency can be negative, and if this negative resistance exceeds the sum of the resistances of the armature and of the network, there will be an overall negative damping at the sub-synchronous frequency, and consequently the sub-synchronous current would increase with time [21], [24]. In SSTI, if the complement of the torsional natural frequency of the drive-train shaft system of the DFIG wind turbine happens to be close to the electric natural frequency of the electrical network, the sub-synchronous torque components generated by the sub-synchronous induced armature voltage can be sustained [21], [24]. The nature of the SSCI is different from SSIGE and SSTI, since in this type of SSR, the DFIG wind turbine controllers play the main role in creating the SSCI; indeed, the SSCI occurs as a results of interaction between the series compensated electrical network and the DFIG wind turbine controllers [27] - [32].

Although the SSR analysis and damping in traditional power systems are well-known and have been extensively studied in the literature [14] - [16], [33], this problem in series-compensated wind farms requires more study and analysis. In particular, after the SSR event that occurred in the Electric Reliability Council of Texas (ERCOT) in 2009 [27] - [31], the wind power industry has become quite interested in SSR studies. In the ERCOT SSR event, a faulted line and subsequent outage in the network caused the radial connection of a large DFIG wind farm to the series-compensated network, resulting in severe sub-synchronous frequency oscillations that caused damage to both the series capacitor and the wind turbine [27] - [31].

A proper modal analysis of a series compensated DFIG-based wind farm can provide control engineers with critical information on the system dynamic characteristics and help them design appropriate SSR mitigation techniques for the system. In this book, we first present a comprehensive approach to modal analysis of a series compensated DFIG-based wind farm using Matlab/Simulink, which will provide the required background for readers to develop the baseline model of the DFIG system for their small-signal stability studies. Afterward, the different types of SSR that may occur in wind farms are presented. The damping methods of the SSR in wind farms will be covered in a subsequent book. The organization of the book is as follows:

Chapter 2 presents modal analysis of a DFIG-based series compensated wind farm using Matlab/Simulink. The model of the system includes wind turbine aerodynamics, a sixth-order induction generator, a second-order two-mass shaft system, a fourth-order series compensated transmission line, a fourth-order rotor-side converter (RSC), a fourth-order grid-side converter (GSC), and a first-order DC-link model. The following three chapters focus on the identification and definition of the main types of the SSR that occur in DFIG wind farms.

Chapter 3 covers the subsynchronous induction generator effect (SSIGE). First a simple definition of the SSIGE is given; then, using eigenvalue analysis and time-domain simulations, it is shown that the DFIG wind farm can be highly unstable due to the SSIGE; finally, the impact of wind speed and compensation level variations on the SSIGE is explained.

Chapter 4 covers subsynchronous torsional interactions (SSTI). First a descriptive definition is given; then, the real world possibility of the SSTI in DFIG wind farm is studied; finally, the impact of the turbine stiffness coefficient and compensation level variations on this type of SSR is investigated.

Chapter 5 briefly describes subsynchronous control interactions (SSCI). Since the SSCI may be confused with the SSIGE, a simple definition of the SSCI is given and a description of how it occurs in DFIG wind farm is provided.

CHAPTER 2

Modeling of DFIG-Based Wind Farm for SSR Analysis

This chapter presents a step-by-step procedure for modal analysis of a DFIG-based series compensated wind farm using MATLAB/Simulink. The model of the system includes wind turbine aerodynamics, a sixth-order induction generator, a second-order two-mass shaft system, a fourth-order series compensated transmission line, a fourth-order rotor-side converter (RSC), a fourth-order grid-side converter (GSC), and a first-order DC-link model.

2.1 POWER SYSTEM DESCRIPTION

The studied power system, shown in Figure. 2.1, is adapted from the IEEE first benchmark model (FBM) for SSR studies [35]. In this system, a 100 MW DFIG-based offshore wind farm is connected to the infinite bus via a 161 kV series compensated transmission line [3], [22], and [36]. The 100 MW wind farm is an aggregated model of 50 wind turbine units, where each unit has a power rating of 2 MW. In fact, a 2 MW wind turbine is scaled up to represent the 100 MW wind farm. This simplification is supported by several studies [37], [38]. The system parameters are given in the Appendix A.

2.2 SMALL-SIGNAL STABILITY

Small-signal stability is the ability of a system to maintain stability when it is subjected to small disturbances [51]. The small-signal stability analysis of a power system can provide power system designers with valuable information about the inherent small-signal dynamic characteristics of the power system, which will help them in the power system design process.

The behavior of a dynamic system, e.g., a power system, can be expressed by a set of n first order nonlinear ordinary differential equations as follows [39]:

$$\dot{x} = f(x, u) \tag{2.1}$$

where $x = [x_1 \ x_2 \ ...x_n]^T$ is the state vector, and its elements, x_i, are the state variables. The column vector $u = [u_1 \ u_2 \ ...u_r]^T$ is the input vector and its elements are the system inputs.

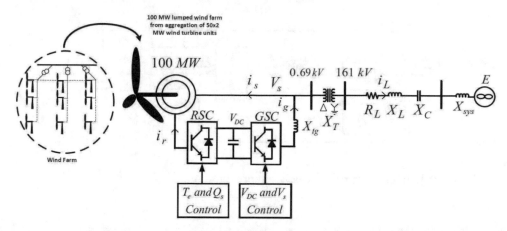

Figure 2.1: One line diagram of the studied power system. R_L = transmission line resistance, X_L = transmission line reactance, X_T = transformer reactance, X_{sys} = system impedance, X_C = fixed series capacitor reactance, X_{tg} = reactance of the grid side converter (GSC) transformer, V_s = generator's terminal voltage, i_L = line current, i_g = GSC current, i_s = stator current, i_r = rotor current. See [3], [22], and [35].

We might also be interested in the output variables, which can be expressed in terms of the state and the input variables as follows [39]:

$$y = g(x, u) \tag{2.2}$$

where $y = [y_1 \ y_2 \ ... y_m]^T$ is the vector of outputs, and $g(x, u)$ is a vector of nonlinear functions that relates the outputs to the inputs and the state variables.

If the disturbances to the system are considered sufficiently small, one can linearize the differential equations 2.1 and 2.2 around an operating point and can express the system dynamics in state-space form as follows:

$$\Delta \dot{x} = A \ \Delta x + B \ \Delta u \tag{2.3}$$
$$\Delta y = C \ \Delta x + D \ \Delta u \tag{2.4}$$

where A is the state matrix of size $n \times n$, B is the input matrix of size $n \times r$, C is the output matrix of size $m \times n$ and D is the feed-forward matrix of size $m \times r$. The generic block diagram representing equations 2.3 and 2.4 is shown in Figure 2.2.

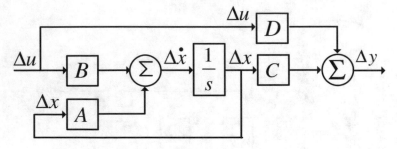

Figure 2.2: Block diagram of the state-space representation.

2.3   TRANSFORMATION FROM *abc* TO *qd0* FRAME

In this work, in order to make the calculations easier, three-phase *abc* variables are transformed into *qd0* variables using the following equation in matrix notation [40].

$$f^e_{qd0s} = K^e_{qd0s} \, f^e_{abcs} \tag{2.5}$$

where

$${f^e_{qd0s}}^T = [f^e_{qs} \; f^e_{ds} \; f^e_{0s}] \tag{2.6}$$

$${f^e_{abcs}}^T = [f^e_{as} \; f^e_{bs} \; f^e_{cs}] \tag{2.7}$$

$$K^e_{qd0s} = \frac{2}{3}
\begin{bmatrix}
\cos\theta & \cos(\theta - \dfrac{2\pi}{3}) & \cos(\theta + \dfrac{2\pi}{3}) \\
\sin\theta & \sin(\theta - \dfrac{2\pi}{3}) & \sin(\theta + \dfrac{2\pi}{3}) \\
\dfrac{1}{2} & \dfrac{1}{2} & \dfrac{1}{2}
\end{bmatrix} \tag{2.8}$$

where f^e denotes voltage, current, flux linkage, or electric charge, θ is the angle between the rotating q-axis and the stationary a-axis, and $\omega_e = \dfrac{d\theta}{dt}$, where ω_e is the rotating synchronous frame frequency.

Figure 2.3 shows the *qd*-frame with respect to the stator *abc*-frame, where *q*-axis is leading the *d*-axis. Note that in this work, the synchronously rotating reference frame is used, in which the reference frame rotates at the electrical angular velocity of the air-gap rotating magnetic field generated by stator currents at the fundamental frequency, ω_e.

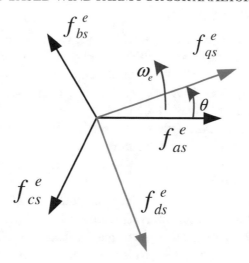

Figure 2.3: Synchronously rotating *qd* frame with respect to the stator *abc*-frame.

2.4 WIND-TURBINE AERODYNAMICS

The torque created by the wind on the turbine shaft can be calculated from the wind speed V_w as follows [22], [41]:

$$T_\omega = \frac{P_\omega}{\omega_m} = \frac{0.5\rho\pi R^2 C_P V_w^2}{\omega_m} \tag{2.9}$$

P_ω is the wind power (W), T_ω is the wind torque $(N \cdot m)$, V_ω is the wind speed (m/s), ρ is the air density $(kg \cdot m^{-3})$, R is the rotor radius of the wind turbine (m), and ω_m is the wind turbine shaft speed (rad/s).

Coefficient C_P is the power coefficient of the blade given by:

$$C_P = 0.5(\frac{RC_f}{\lambda_w} - 0.022\theta - 2)e^{-0.225\frac{RC_f}{\lambda_\omega}} \tag{2.10}$$

where C_f is the wind turbine blade design constant, and θ is the wind speed pitch angle (rad)

Also, λ_ω is the wind speed tip-speed ratio defined by:

$$\lambda_\omega = \frac{\omega_m R}{V_\omega} \tag{2.11}$$

2.5 DFIG CONVERTER CONTROL METHODS

In this section, a short overview of rotor-side converter (RSC) and grid-side converter (GSC) controllers is presented.

2.5.1   RSC CONTROLLER

Usually, the RSC is used to control the electric torque (or the rotor speed) of the DFIG and to control the power factor at the stator terminals [42]. Different control strategies for the RSC, including vector control (VC) [43] - [45], direct torque control (DTC) [46] - [48], and direct power control (DPC) [49] - [51] have been reported in the literature.

In the VC method, the rotor currents are usually controlled using a *qd* rotating frame aligned with the stator flux. In this frame the electric torque is proportional to the q-axis rotor current. The VC method controls the electric torque by controlling this rotor current component. Moreover, reactive power control in the machine is achieved by controlling the d-axis component of the rotor current [43] - [45].

In the DTC method, the rotor flux linkage magnitude and the electric torque of the DFIG are directly controlled. This direct control is accomplished by proper control of the inverter on the rotor side. The DTC method utilizes flux and torque feedback loops. The machine flux is estimated using the rotor and stator current vectors while the machine electric torque is estimated using the estimated rotor flux and the measured rotor currents.

The DPC method is similar to the DTC, but it considers the effect of both the stator and rotor fluxes upon the real and reactive power of the stator. Indeed, this method aims to directly control the real and reactive power of the stator by applying the appropriate rotor voltage vector to the machine [48] - [51].

2.5.2   GSC CONTROLLER

The main objective of the GSC is to regulate the DC-link voltage and to permit real power flow through the converter. For the GSC, the VC method is usually adopted, see [3], [22], and [52], where the reference frame is aligned with the grid-voltage vector. Additionally, DPC method has been implemented in the literature to control the GSC, resulting in independent real and reactive power flow in the converter [53].

2.5.3   MODELING OF THE DFIG CONVERTER CONTROLLERS

Control loops for RSC and GSC presented in [3], [22], and [54] - [56] are considered in this work. Both RSC and GSC controllers are modeled. In order to maximize power generation in the DFIG wind farm, the maximum power point tracking (MPPT) method is used [54]. Figure 2.4 shows the wind power versus wind turbine shaft speed in per unit for various wind speeds with indication of MPPT curve. To enforce operation on the MPPT curve, for a given wind speed V_ω, the optimal reference power and optimal rotational speed obtained from Figure 2.4 are used to calculate the reference torque, T_e^*. Note that due to power converter ratings, it may not be practical to always work on the MPPT curve. On the one hand, for very low wind speeds, the DFIG operates at almost constant rotational speed. On the other hand, when the wind speed increases so that the MPPT torque reference exceeds the turbine torque rating, the DFIG is operated at maximum constant torque [57].

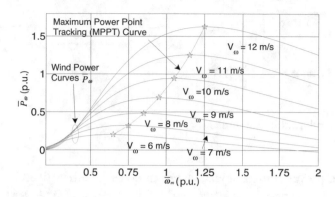

Figure 2.4: Wind power $\bar{P}_m$ (*p.u.*), wind turbine shaft speed $\bar{\omega}_m$ (*p.u.*), and wind speed V_ω (*m/s*) relationship [3] and [22].

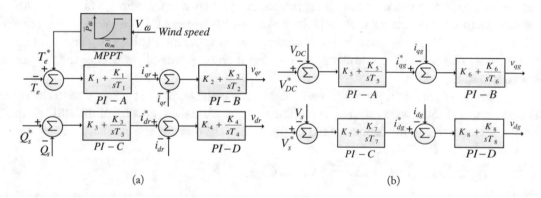

Figure 2.5: (a) RSC controllers. (b) GSC controllers [22].

The aim of the GSC and RSC is to control the DFIG so that in steady state it operates on the MPPT curve. Figure 2.5 shows the block diagrams of the two controllers. The GSC controller is responsible for controlling the DC-link voltage, V_{DC}, and the induction generator's terminal voltage, V_s [22] and [54]. The RSC controller is responsible for regulating the electric torque, T_e, and stator reactive power, Q_s. In steady state condition, for operation on the MPPT curve, neglecting power losses, the electric torque, T_e, must be equal to the wind torque,

$$T_\omega = \frac{P_\omega}{\omega_m} \qquad (2.12)$$

Therefore, the reference torque, T_e^*, can be calculated based on the value of T_ω^* obtained from the MPPT curve shown in Figure 2.5. The value of reference reactive power Q_s^* depends on

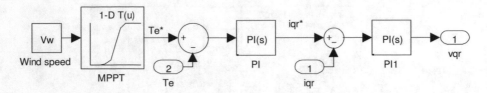

Figure 2.6: GSC regulators in Simulink.

the chosen reactive power control objective, which could be either fixed reactive power or unity power factor [54]. In this paper, the latter method is chosen and Q^* is set to zero.

Note that the grid-side and rotor-side converters are considered to be ideal: transparent to power and having infinite bandwidth. They are transparent to power in the sense that they are assumed to store no energy and their internal losses are neglected. They have infinite bandwidth in the sense that, given the high switching frequency, they operate fast enough in comparison with other system time constants that their dynamics can be neglected. Consequently, the outputs of the RSC controller are directly the q and d components of the rotor voltage, v_{qr} and v_{dr}. Similarly, the outputs of the GSC controller are the q and d components of the voltage at the grid-side transformer, v_{qg} and v_{dg}.

The GSC and RSC controllers add eight state variables to the system, due to the eight PI controllers, and their state variables are defined as a vector X_{RG}. One loop of the RSC controllers implemented in Simulink is represented in Figure 2.6. The other controllers have similar structure.

2.6   MODELING OF THE DC-LINK

In this work, the DC link capacitor dynamics are modeled. Figure 2.7 shows the back-to-back converters between DFIG and grid which share a common DC-link. The dynamics of the DC-link can be expressed by a first order nonlinear model as follows [3], [22], and [58]:

$$- C v_{DC} \frac{dv_{DC}}{dt} = P_r + P_g \tag{2.13}$$

where the rotor side converter's (RSC) active power P_r, and the grid-side converter's (GSC) active power, P_g, are given as follows [40]:

$$P_r = 0.5 \left(v_{qr} \, i_{qr} + v_{dr} \, i_{dr} \right) \tag{2.14}$$

$$P_g = 0.5 \left(v_{qg} \, i_{qg} + v_{dg} \, i_{dg} \right) \tag{2.15}$$

Implementation of the DC-link model in Matlab/Simulink is shown in Figure 2.8. Notice that P_r and P_g are added together and then multiplied by the base power, S_{base}, in order to obtain the power in MW. This is necessary, since per unit values are not used for the DC-link model, and all values in this block are actual physical values.

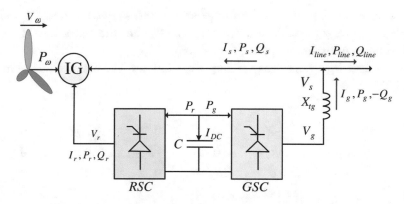

Figure 2.7: Back-to-back converter between the DFIG and grid.

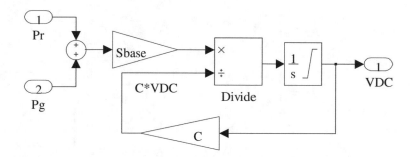

Figure 2.8: DC-link model in Matlab/Simulink. In this figure, $S_{basae} = 100\,MW$ and $C = 50 * 14000$ μF.

2.7 MODELING OF THE INDUCTION MACHINE

Using the *abc-qd0* transformation described in Section 2.3, the induction machine equations in *abc*-frame are transformed into the synchronously rotating *qd*-frame. If the DFIG currents are selected as the state variables, then the DFIG *qd* model in per unit will be as follows, where the input variables are the DFIG voltages [22] and [40]:

$$\dot{X}_{DFIG} = A_{DFIG}X_{DFIG} + B_{DFIG}U_{DFIG} \tag{2.16}$$

where

$$X_{DFIG} = [i_{qs}\ i_{ds}\ i_{0s}\ i_{qr}\ i_{dr}\ i_{0r}]^T \tag{2.17}$$

$$U_{DFIG} = [v_{qs} \; v_{ds} \; v_{0s} \; v_{qr} \; v_{dr} \; v_{0r}]^T \tag{2.18}$$

where $i_{qs}, i_{ds}, i_{qr}, i_{dr}$ are the stator and rotor qd-axis currents ($p.u.$), $v_{qs}, v_{ds}, v_{qr}, v_{dr}$ are the stator and rotor qd-axis voltages ($p.u.$), and $i_{0s}, i_{0r}, v_{0s}, v_{0r}$ are the stator and rotor zero sequence current and voltage components ($p.u.$), respectively.

The A_{DFIG} and B_{DFIG} matrices are defined as follows. We first define the matrices given in Eq. 2.19 and Eq. 2.20.

$$F = \begin{bmatrix} R_s & \frac{\omega_e}{\omega_b} X_{SS} & 0 & 0 & \frac{\omega_e}{\omega_b} X_M & 0 \\ -\frac{\omega_e}{\omega_b} X_{SS} & R_s & 0 & -\frac{\omega_e}{\omega_b} X_M & 0 & 0 \\ 0 & 0 & R_s & 0 & 0 & 0 \\ 0 & \frac{(\omega_e - \omega_r)}{\omega_b} X_M & 0 & R_r & 0 & \frac{(\omega_e - \omega_r)}{\omega_b} X_{rr} \\ -\frac{(\omega_e - \omega_r)}{\omega_b} X_M & 0 & 0 & -\frac{(\omega_e - \omega_r)}{\omega_b} X_{rr} & R_r & 0 \\ 0 & 0 & 0 & 0 & 0 & R_r \end{bmatrix} \tag{2.19}$$

$$G = \begin{bmatrix} X_{ss} & 0 & 0 & X_M & 0 & 0 \\ 0 & X_{ss} & 0 & 0 & X_M & 0 \\ 0 & 0 & X_{ls} & 0 & 0 & 0 \\ X_M & 0 & 0 & X_{rr} & 0 & 0 \\ 0 & X_M & 0 & 0 & X_{rr} & 0 \\ 0 & 0 & 0 & 0 & 0 & X_{lr} \end{bmatrix} \tag{2.20}$$

Then:

$$A_{DFIG} = -\omega_b \cdot G^{-1} \cdot F \tag{2.21}$$

$$B_{DFIG} = \omega_b \cdot G^{-1} \tag{2.22}$$

In Eq. 2.19 and 2.20: X_{lr} is the rotor leakage reactance ($p.u.$), X_{ls} is the stator leakage reactance ($p.u.$), X_M is the magnetizing reactance ($p.u.$), the stator reactance X_{ss} is $X_{ss} = X_{ls} + X_M$ ($p.u.$), the rotor reactance X_{rr} is $X_{rr} = X_{lr} + X_M$ ($p.u.$), R_r is the rotor resistance ($p.u.$), R_s is the stator resistance ($p.u.$), ω_b is the base frequency (rad/s), ω_r is the generator rotor speed (rad/s), and ω_e is the rotating synchronous frame frequency (rad/s).

The DFIG dynamic model of Eq. 2.16 is implemented in Matlab/Simulink. The simulated system is given in Figure 2.9. As seen in this figure, the inputs to the system are the DFIG qd-frame stator and rotor voltages. Rotor voltages come from the RSC controller of Figure 2.5 and stator voltages are given by equation 2.39 later in this chapter. The state variables are the DFIG

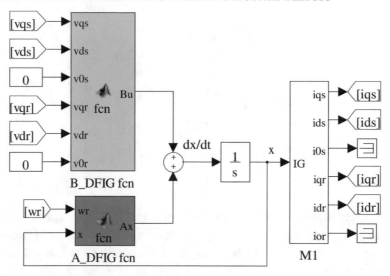

Figure 2.9: DFIG model in Matlab/Simulink.

qd-frame stator and rotor currents, as shown in Figure 2.9. Note that since we are considering a balanced system, the stator and rotor zero-sequence voltage components, i.e., v_{0s} and v_{0r}, are set to zero. Also, note that the input ω_r to the system is the generator rotor speed in (rad/s), which is provided by the shaft equations, which will be given in the next subsection.

2.8 MODELING OF THE SHAFT SYSTEM

The shaft of the wind turbine system can be represented by a two-mass system. The first mass represents the low-speed turbine and the second mass represents the high-speed generator, and the two mass connection is modeled as a spring and a damper. The motion equations can be expressed as 3 first order differential equations in per unit as follows, see [22] and [58]:

$$\dot{X}_{shaft} = A_{shaft} X_{shaft} + B_{shaft} U_{shaft} \tag{2.23}$$

where

$$X_{shaft} = [\bar{\omega}_m \ \bar{\omega}_r \ T_{tg}]^T \tag{2.24}$$

$$U_{shaft} = [\bar{T}_\omega \ T_e \ 0]^T \tag{2.25}$$

The A_{shaft} and B_{shaft} matrices are defined as follows:

$$A_{shaft} = \begin{bmatrix} \frac{(-D_t - D_{tg})}{2H_t} & \frac{D_{tg}}{2H_t} & \frac{-1}{2H_t} \\ \frac{D_{tg}}{2H_g} & \frac{(-D_t - D_{tg})}{2H_g} & \frac{-1}{2H_g} \\ K_{tg}\omega_b & -K_{tg}\omega_b & 0 \end{bmatrix} \qquad (2.26)$$

$$B_{shaft} = \begin{bmatrix} \frac{1}{2H_t} & 0 & 0 \\ 0 & \frac{1}{2H_g} & 0 \\ 0 & 0 & 1 \end{bmatrix} \qquad (2.27)$$

In the shaft equations, $\bar{\omega}_m$ is the turbine shaft speed ($p.u.$), $\bar{\omega}_r$ is the generator rotor speed ($p.u.$), $\bar{T}_\omega$ is the wind torque ($p.u.$), D_g and D_t are damping coefficient of generator and turbine ($p.u.$), D_{tg} is the damping coefficient between the two masses ($p.u.$), K_{tg} is the inertia constant of turbine and generator ($p.u/rad$), and H_g and H_t are the inertia constants of generator and turbine (s).

For the shaft equations, the state variables are wind turbine speed $\bar{\omega}_m$, generator rotor speed $\bar{\omega}_r$, and internal torque of the two-mass system T_{tg}. The inputs to the two-mass model are wind torque T_ω and electric torque T_e. The optimal value of the wind torque for any given wind speed can be obtained using the MPPT curve shown in Figure 2.4. The value of the electric torque can be calculated using the following equations [22] and [40]:

$$\Psi_{qm} = X_m(i_{qs} + i_{qr}) \qquad (2.28)$$

$$\Psi_{dm} = X_m(i_{ds} + i_{dr}) \qquad (2.29)$$

$$T_e = 0.5(\Psi_{qm}i_{dr} - \Psi_{dm}i_{qr}) \qquad (2.30)$$

Figure 2.10 represents the model implementation of the shaft system in Matlab/Simulink. As seen in this figure, the DFIG stator currents, i_{qs}, i_{ds}, and rotor currents, i_{qr}, i_{dr}, are applied to the Matlab Function (fcn), T_e, to calculate electric torque using Eqs. 2.28 - 2.30. Then Eq. 2.23 is implemented using the fcn blocks A_{shaft} and B_{shaft} and an integrator block $\frac{1}{s}$ to implement the shaft model. The state variable ω_r, calculated by the shaft model of Figure 2.10, is provided to the DFIG model of Figure 2.9.

2.9   MODELING OF THE TRANSMISSION LINE

The transformation described in Section 2.3 is used to convert the transmission line equations from *abc*-frame to *qd*-frame. Considering the line current and the voltage across the capacitor

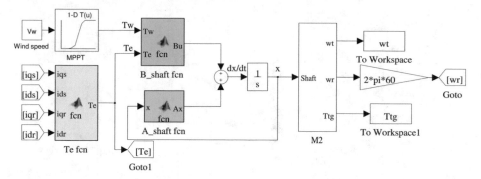

Figure 2.10: The shaft system model in Matlab/Simulink.

as state variables, the transmission line equations in qd-frame can be expressed in matrix form as follows, see [4] and [40]:

$$\dot{X}_{Tline} = A_{Tline}X_{Tline} + B_{Tline}U_{Tline} \tag{2.31}$$

where

$$X_{Tline} = [i_{ql}\ i_{dl}\ v_{qc}\ v_{dc}]^T \tag{2.32}$$

$$U_{Tline} = [\frac{(v_{qs} - E_{Bq})}{X_L}\ \frac{(v_{ds} - E_{Bd})}{X_L}\ 0\ 0]^T \tag{2.33}$$

The A_{Tline} and B_{Tline} matrices are defined as follows:

$$A_{Tline} = \begin{bmatrix} \frac{-R_L}{X_L} & -\bar{\omega}_e & \frac{-1}{X_L} & 0 \\[2mm] \bar{\omega}_e & \frac{-R_L}{X_L} & 0 & \frac{-1}{X_L} \\[2mm] X_C & 0 & -\bar{\omega}_e & 0 \\[2mm] 0 & X_C & \bar{\omega}_e & 0 \end{bmatrix} \tag{2.34}$$

$$B_{Tline} = \begin{bmatrix} \omega_b & 0 & 0 & 0 \\[1mm] 0 & \omega_b & 0 & 0 \\[1mm] 0 & 0 & 1 & 0 \\[1mm] 0 & 0 & 0 & 1 \end{bmatrix} \tag{2.35}$$

where i_{ql} and i_{dl} are the transmission line qd-axis currents ($p.u.$), v_{qc} and v_{dc} are the series capacitor's qd-axis voltages ($p.u.$), R_L is the transmission line resistance ($p.u.$), X_L is the transmission

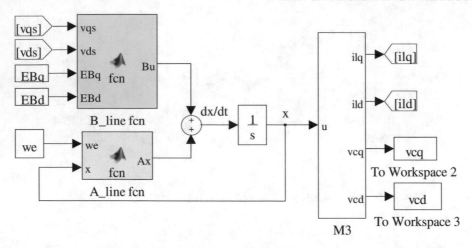

Figure 2.11: Transmission line model in Matlab/Simulink.

line reactance ($p.u.$), X_C is the fixed series capacitor impedance ($p.u.$), E_{Bq} and E_{Bd} are the infinite bus qd-axis voltages ($p.u.$), and $\bar{\omega}_e$ is the rotating synchronous frame frequency ($p.u.$)

Since we are dealing with a three-phase balanced system, the zero sequence components can be neglected. Whereupon, it can be observed from these equations that the state variables in the transmission line system are i_{lq}, i_{ld}, v_{cq}, and v_{cd}. The Simulink model of the transmission line in a synchronously rotating qd-frame is shown in Figure 2.11.

2.10 INTEGRATING THE MODELS

So far dynamic equations of the various subsystems shown in Figure 2.1 have been presented. However, some additional algebraic equations are needed to interface these dynamic equations and obtain the complete system model. These algebraic equations are described in this section. By applying KCL at the common point of the stator, GSC, and transmission line (see Figure 2.7), the following equation can be obtained:

$$I_g = I_s + I_{line} \tag{2.36}$$

This equation expressed in $qd0$ coordinates yields the two equations:

$$i_{qg} = i_{qs} + i_{ql} \quad i_{dg} = i_{ds} + i_{dl} \tag{2.37}$$

In this work, the transformer in GSC-side of the generator is considered lossless, and its dynamics are neglected. Consequently the transformer can be modeled by an inductive impedance X_{tg}. Applying KVL between the GSC output and the stator of the DFIG (see Figure 2.7) one obtains the following phasor equation:

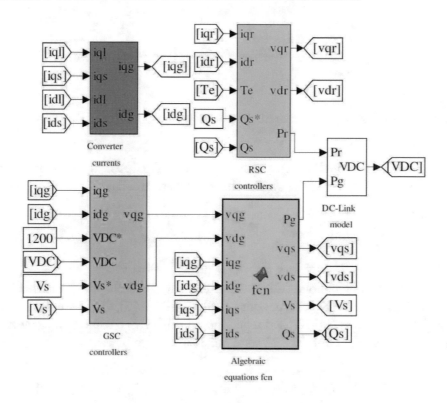

Figure 2.12: RSC and GSC controllers, DC-link, and algebraic equationss in Matlab/Simulink.

$$V_g - V_s = jX_{tg} I_g \tag{2.38}$$

Expressing this equation in qd coordinates results in:

$$v_{qs} = v_{qg} - X_{tg} i_{dg} \quad v_{ds} = v_{dg} + X_{tg} i_{qg} \tag{2.39}$$

The stator reactive power, Q_s, used in the RSC controller block of Figure 2.5, is given by:

$$Q_s = 0.5(v_{qs}i_{ds} - v_{ds}i_{qs}) \tag{2.40}$$

Figure 2.12 shows the GSC and RSC controllers and DC-link model in Matlab/Simulink. The mentioned algebraic equations, described in this section, are implemented in the "Converter currents" block and "Algebraic equations" fcn of Figure 2.13. Equation 2.37 is implemented in the "Converter Currents" block. Equations (2.15) and (2.39) are implemented in the "Algebraic equations fcn" block.

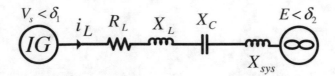

Figure 2.13: The system equivalent circuit in steady state.

Considering the modeling of the system shown in Figure 2.1 described in this chapter, the entire DFIG system is a 22nd order system that can be expressed as:

$$\dot{X} = f(X, U, t) \tag{2.41}$$

where

$$X = [X_{IG}^T \parallel X_{shaft}^T \parallel X_{Tline}^T \parallel v_{DC} \parallel X_{RG}^T]^T \tag{2.42}$$

This system model combines all models described in this chapter. Note that this system of ODEs is nonlinear. For example, equations (2.13), (2.14), (2.15), (2.30) are nonlinear. The entire Simulink system is named "DFIG" and represents a DFIG-based wind farm connected to a series compensated transmission line. The 22 state variables are as follows:

$$\begin{aligned} X = [&i_{qs}\ i_{ds}\ i_{0s}\ i_{qr}\ i_{dr}\ i_{0r} \parallel \bar{\omega}_m\ \bar{\omega}_r\ T_{tg} \parallel i_{ql}\ i_{dl}\ v_{qc}\ v_{dc} \parallel v_{DC} \parallel \\ &i_{qr}^*\ i_{dr}^*\ v_{qr}\ v_{dr}\ i_{qg}^*\ i_{dg}^*\ v_{qg}\ v_{dg}]^T \end{aligned} \tag{2.43}$$

Out of 22 state variables, 3 state variables are related to the shaft system, 6 state variables belong to the DFIG model, 4 state variables are related to the transmission line, 1 state variable is for the DC-link, and 8 state variables model the RSC and GSC controllers.

2.11 MODAL ANALYSIS

Given a Simulink system of nonlinear differential equations such as the "DFIG" system described above, Matlab/Simulink can extract the state-space linearized model evaluated at a given operating point. Different linearization algorithms are available in different versions of Matlab/Simulink. In general these algorithms either combine together precomputed analytical Jacobians for each Simulink block or perform a small-signal numerical perturbation around the specified operating point [59]. In some cases a combination of these two approaches is used. See also the documentation for the Matlab command LINMOD. The linearization procedure provides the state-space linearized matrices A, B, C and D shown in Figure 2.2. However, in order to linearize the system, the operating-point values of the state variables must be provided to the Matlab/Simulink. The next subsection describes the procedure used to determine the values of the state variables at the desired operating point.

2.11.1   OBTAINING THE STEADY-STATE OPERATING POINT

The objective of this procedure is to find the state vector X_0 at the steady-state operating point.

$$X_0 = [i_{qs0}\ i_{ds0}\ i_{0s0}\ i_{qr0}\ i_{dr0}\ i_{i0r0} \parallel \bar{\omega}_{m0}\ \bar{\omega}_{r0}\ T_{tg0} \parallel i_{ql0}\ i_{dl0}\ v_{qc0}\ v_{dc0} \parallel v_{DC0} \parallel$$
$$i_{qr0}^*\ i_{dr0}^*\ v_{qr0}\ v_{dr0}\ i_{qg0}^*\ i_{dg0}^*\ v_{qg0}\ v_{dg0}]^T \tag{2.44}$$

The steady-state operating point is evaluated for a given compensation level K and for a given value of wind speed V_ω. The following values will be used to illustrate the procedure: K=75% and V_ω = 7 m/s. From the MPPT curve of Figure 2.4 we can find the desired wind power P_ω^* = 0.322 p.u. and the desired shaft speed ω_m =0.75 p.u. for the given wind speed. The reference torque for the RSC controller of Figure 2.5 is given by:

$$T_e^* = \frac{P_\omega^*}{\omega_m} = \frac{0.322}{0.75} = 0.43\ p.u. \tag{2.45}$$

Furthermore, we assume that the infinite bus voltage and the DFIG stator voltage are at nominal values, i.e., $E = V_s = 1$ p.u .

The system steady-state vector X_0 is obtained in two steps. In step 1, a load-flow analysis is carried out on the DFIG connected to the infinite bus considering the DFIG stator node as a voltage-controlled (PV) bus. This is done to obtain the angle between the DFIG bus voltage and the infinite bus voltage. Then in step 2, using the obtained solution, the initial values of the state variables can be calculated by solving the state equations, with time derivative equal to zero, and control outputs equal to their reference values.

For step 1, the system under study, shown in Figure 2.1, can be simplified for load flow analysis as shown in Figure 2.13. The angle $\delta = \delta_1 - \delta_2$ is the angle of the DFIG terminal voltage with respect to the infinite bus voltage. With P_ω^*, V_s and E as known variables, the angle δ is obtained from the load-flow analysis equation (2.46).

$$\frac{R_L V_s^2}{R_L^2 + (X_L + X_{sys} - X_C)^2} - \frac{E V_s R_L}{R_L^2 + (X_L + X_{sys} - X_C)^2} \cos\delta$$
$$+ \frac{E V_s (X_L + X_{sys} - X_C)}{R_L^2 + (X_L + X_{sys} - X_C)^2} \sin\delta - P_\omega^* = 0 \tag{2.46}$$

The DFIG stator voltage is assumed to be aligned with the q-axis of the dq-frame, as shown in Figure 2.14. Consequently, $v_{ds} = 0$ and $v_{qs} = V_s = 1$ p.u.

Notice that v_{ds} and v_{qs} are not state variables, since they can be calculated from Equation (2.39) as a function of state variables. The infinite bus voltage components in dq-frame are given by:

$$E_d = E_B \sin\delta \tag{2.47}$$

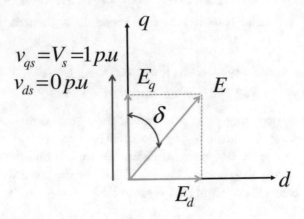

Figure 2.14: Block diagram of the state-space representation.

$$E_q = E_B \cos \delta \qquad (2.48)$$

with $E_S = E = 1$.

In step 2, the values of the 22 components of the state vector X_0 are calculated. The DC link voltage is assumed to be at its nominal value

$$v_{DC_0} = 1200 \; V \qquad (2.49)$$

For the shaft system it is assumed that the two masses are rotating together at the desired shaft speed with the desired wind-generated torque. Therefore

$$\omega_{m0} = \omega_{r0} = 0.75 \; p.u \qquad (2.50)$$

$$T_{tg0} = T_e^* = 0.43 \; p.u \qquad (2.51)$$

The zero-sequence components of the stator and rotor currents are set to zero.

$$i_{0s0} = i_{0r0} = 0 \qquad (2.52)$$

The RSC and GSC controller model needs to be treated in a special way. Given the steady-state condition, the eight error terms at the input of the PI controllers in Figure 2.5 are all assumed to be zero with the integral terms providing the appropriate steady-state values at their outputs. For example, we assume $T_e = T_e^*$ so that the input of PI-A is zero and its output is

$$i_{qr0}^* = i_{qr0} \qquad (2.53)$$

Therefore, we cannot use the RSC and GSC controller models of Figure 2.5 to determine the DFIG machine stator and rotor voltages v_{dr0}, v_{qr0}, v_{dg0}, and v_{qg0}. Instead we can write real and reactive power balance equations for the DFIG machine and for the node connecting the DFIG stator to the transmission line and to the transformer reactance X_{tg}. Referring to Figure 2.7 the DFIG reactive power balance equation is given by (2.54) and the DFIG real power balance by (2.55). Note that for the case of unity power factor control it is $Q_s^* = 0$.

$$-0.5(v_{qs0}i_{ds0} - v_{ds0}i_{qs0}) - Q_s^* = 0 \qquad (2.54)$$

$$\begin{aligned} P_{s0} + P_{r0} + P_\omega^* &= 0.5(v_{qs0}i_{qs0} + v_{ds0}i_{ds0}) + 0.5(v_{qr0}i_{qr0} + v_{dr0}i_{dr0}) + P_\omega^* \\ &= 0 \end{aligned} \qquad (2.55)$$

Similarly, Equations (2.56) and (2.57) are the real and reactive power balance equations at the node connecting the DFIG stator to the transmission line and to the transformer reactance X_{tg}, respectively.

Similarly, the real and reactive power balance equations are (2.56) and (2.57), respectively.

$$\begin{aligned} P_{L0} + P_{s0} - P_{g0} &= 0.5(v_{qs0}i_{ql0} + v_{ds0}i_{dl0}) + 0.5(v_{qs0}i_{qs0} + v_{ds0}i_{ds0}) \\ &\quad - 0.5(v_{qg0}i_{qg0} + v_{dg0}i_{dg0}) \\ &= 0 \end{aligned} \qquad (2.56)$$

$$\begin{aligned} Q_{L0} + Q_{s0} + Q_{g0} &= 0.5(v_{qs0}i_{dl0} - v_{ds0}i_{ql0}) + 0.5(v_{qs0}i_{ds0} - v_{ds0}i_{qs0}) \\ &\quad + 0.5(v_{qg0}i_{dg0} - v_{dg0}i_{qg0}) \\ &= 0 \end{aligned} \qquad (2.57)$$

The four Equations (2.54)—(2.57) are combined with eight more algebraic equations from the transmission line model and from the induction machine model. In particular, four algebraic equations are obtained from the transmission line equations (2.31) by setting the time derivatives to zero. The final four equations come from the induction machine model equations (2.16), again by setting the time derivatives to zero. The two zero-sequence equations are discarded, since the steady-state values have already been obtained. We obtain a system of 12 nonlinear algebraic equations that can be solved using the Matlab command FSOLVE. In this way we obtain the 12 steady-state variables:

$$X_0 = [i_{qs0}\ i_{ds0}\ i_{qr0}\ i_{dr0}\ i_{ql0}\ i_{dl0}\ v_{qc0}\ v_{dc0}\ v_{qr0}\ v_{dr0}\ v_{qg0}\ v_{dg0}]^T \qquad (2.58)$$

Table 2.1: Steady-state values of the system state variables for $K = 75\%$ and $V_\omega = 7\ m/s$

i_{qs0}	−0.6064	i_{ds0}	0.0082	i_{qr0}	0.6206	i_{dr0}	0.3500
ω_{m0}	0.75	ω_{r0}	0.75	T_{tg0}	0.43	v_{DC}	1200
i_{ql0}	0.4525	i_{dl0}	−0.0058	v_{qc0}	0.0022	v_{dc0}	0.1742
i_{qr0}^{*}	0.6206	i_{dr0}^{*}	0.3500	v_{qr0}	0.3661	v_{dr0}	−0.0275
i_{qg0}^{*}	−0.1538	i_{dg0}^{*}	0.0024	v_{qg0}	1.4135	v_{dg0}	−0.0445

Table 2.2: Eigenvalue analysis of the system at 75% series compensation and 7 m/s wind speed

Mode	Eigenvalue	Mode	Eigenvalue
$\lambda_{1,2}$	$0.1287 \pm j131.9136$	λ_{15}	−97.8425
$\lambda_{3,4}$	$-5.2066 \pm j617.1976$	λ_{16}	−0.5000
$\lambda_{5,6}$	$-9.9111 \pm j99.9693$	λ_{17}	−0.0143
$\lambda_{7,8}$	$-0.9221 \pm j5.9992$	λ_{18}	−0.0027
$\lambda_{9,10}$	$-875.4284 \pm j4217.8692$	λ_{19}	−19.9297
$\lambda_{11,12}$	$-0.1159 \pm j0.5247$	λ_{20}	−20.8000
λ_{13}	−2151.3685	λ_{21}	−0.0000
λ_{14}	−114.7491	λ_{22}	−0.0000

So far we have obtained 18 of the 22 state variables in steady state. The remaining four state variables are the outputs of the PI-A and PI-C controllers. Their values are

$$i_{qr0}^{*} = i_{qr0} \tag{2.59}$$

$$i_{dr0}^{*} = i_{dr0} \tag{2.60}$$

$$i_{qg0}^{*} = i_{qg0} = \frac{v_{ds0} - v_{dg0}}{X_{tg}} \tag{2.61}$$

$$i_{dg0}^{*} = i_{dg0} = \frac{v_{qs0} - v_{qg0}}{X_{tg}} \tag{2.62}$$

The last two equations are obtained using (2.39). Table 2.1 summarizes the initial values obtained for the desired operating point.

2.11.2 CALCULATION OF THE EIGENVALUES

The nonlinear 22nd order Matlab/Simulink model can be linearized around the operating point of Table 2.1 and the eigenvalues can be obtained using the following two Matlab commands.:

$$<< \begin{bmatrix} A & B & C & D \end{bmatrix} = linmod\ ('DFIG');\ Eigenvalues = eig(A); \tag{2.63}$$

Table 2.2 shows the list of system eigenvalues when the wind speed is 7 m/s and the compensation level is 75%. The nature of each eigenvalue given in Table 2.2 will be discussed in the next chapters.

SUMMARY

This chapter has presented a step-by-step comprehensive procedure for modal analysis of a series compensated DFIG-based wind farm in Matlab/Simulink. A $6th$ order model has been used for the DFIG including stator and rotor dynamics, and a $3rd$, $4th$, and $1st$ order models have been applied for the drive train two-mass model of the shaft system, series compensated transmission line, and the DC-link, respectively. Also, the dynamics of the both grid-side converter (GSC) and rotor-side converter (RSC) controllers have been considered, which adds 8 more state variables. The presented models have been described together with the corresponding Simulink blocks in order to help the reader understand the modeling process. The procedure to obtain the initial values of the system under study is described in detail, since it is essential for small-signal linearized analysis.

CHAPTER 3

Induction Generator Effect

This chapter focuses on the induction generator effect (SSIGE or simply IGE). First a brief review of series compensation is provided. Then, a simple definition of the SSIGE is given. Using eigenvalue analysis and time-domain simulations, it is shown that the DFIG wind farm can be highly unstable due to the SSIGE. Finally, the impact of wind speed and compensation level variations on the SSIGE is explained.

3.1 SERIES COMPENSATION BASICS

In order to briefly explain series compensation, a simple lossless two-machine system is considered, where the sending-point and receiving-point voltages are assumed to have the same magnitude, as shown in Figure 3.1(A). In this figure, the effective transmission line impedance including the series capacitor is given by:

$$X_{Leff} = X_L - X_C \tag{3.1}$$

The series compensation level (also-called the degree of series compensation) K is defined as:

$$K = \frac{X_C}{X_L} \quad 0 \le K < 1 \quad or \quad 0\% \le K < 100\% \tag{3.2}$$

Substitution of Eq. 3.2 into Eq. 3.1 results in:

$$X_{Leff} = (1 - K)X_L \tag{3.3}$$

Considering the two-machine system of Figure 3.1(A), with the assumption that $V_{send} = V_{res} = V$, it can be easily shown (see [12]) that the line current and real power can be expressed as:

$$I_L = \frac{2V}{(1-K)X_L} \sin\frac{\delta}{2} \tag{3.4}$$

$$P = \frac{P_{max}}{(1-K)} \sin\delta \tag{3.5}$$

where δ is the angle between the V_{send} and V_{rec} phasors, and P_{max} is defined as:

$$P_{max} = \frac{V^2}{X_L} \tag{3.6}$$

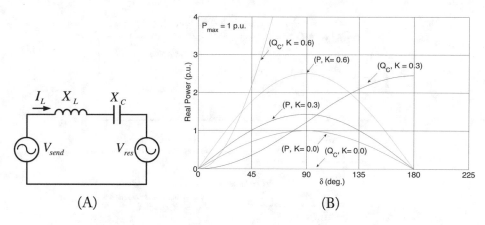

Figure 3.1: (A) A simple lossless series compensated two-machine system. (B) Plot of transmitted real power and reactive power injected by series capacitor as a function of phase angle δ between V_{send} and V_{res}, for different series compensation levels K.

Additionally, the reactive power injected into the line by the series capacitor can be expressed as [12]:

$$Q_C = 2P_{max}\frac{K}{(1-K)^2}(1-\cos\delta) \tag{3.7}$$

Figure 3.1(B) shows real power P and reactive power Q_C versus δ, for different values of series compensation level K. In this figure, it is assumed that P_{max} is equal to 1 $p.u.$. It can be observed that the transmissible real power P of the transmission line increases, as expected from Eq. 3.5, as the series compensation level K increases. Likewise, the injected reactive power by the series capacitor Q_C increases, as K increases. In conclusion, the basic idea of series compensation is to cancel out a portion of the inductive impedance of a transmission line using the capacitive impedance of the series capacitor. This reduces the total inductive reactance of the transmission line, as if the line had been physically shortened. A drawback is that the series capacitor introduces a resonance in the system with resonant frequency equal to the frequency at which the inductive impedance of the transmission line is equal in magnitude to the capacitive impedance of the series compensation capacitor.

3.2   INDUCTION GENERATOR EFFECT (SSIGE)

The general expression of the stator current in a series compensated wind turbine generator system (WTGS) can be defined as [60]:

$$i_L(t) = A\sin(\omega_s t + \phi_1) + Be^{-\alpha t}\sin(\omega_n t + \phi_2) \tag{3.8}$$

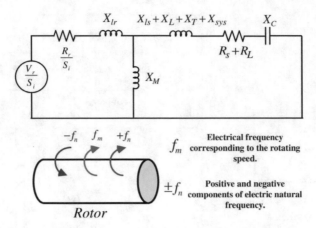

Figure 3.2: Equivalent circuit of the system at sub-synchronous and super-synchronous frequencies.

where ω_s is the electric fundamental frequency and ω_n is the natural frequency of the electric network, which is given by [60]:

$$\frac{\omega_n}{2\pi} = f_n = f_s \sqrt{\frac{KX_e}{\sum X}} \tag{3.9}$$

where $K = \frac{X_C}{X_e}$ is the compensation level, $X_e = X_L + X_T$ $(p.u.)$, $\sum X$ is the entire inductive reactance seen from the infinite bus $(p.u.)$, f_n is the natural frequency of the electric system (Hz), and f_s is the frequency of the system (Hz).

Figure 3.2 shows the equivalent circuit of the DFIG wind turbine at sub-synchronous and at super-synchronous frequencies. This figure also shows the rotor with the electrical frequency corresponding to the rotating speed and the positive and negative components of the electric natural frequency. At sub-synchronous and super synchronous frequencies, the slip is given by S_1 and S_2, respectively, as follows:

$$S_1 = \frac{f_n - f_m}{f_n}, \quad S_2 = \frac{f_n + f_m}{f_n} \tag{3.10}$$

The super-synchronous slip, i.e., S_2 in Eq. 3.10, is always a positive value, and consequently, $\frac{R_r}{S_2}$ in Figure 3.2 is a positive value. Thus, the DFIG wind farm is stable at this frequency. On the other hand, the sub-synchronous slip, i.e., S_1 in Eq. 3.10, is a negative number since the electric natural frequency, f_n, is less than the electric frequency corresponding to the rotating speed, f_m. If the magnitude of the equivalent rotor resistance, i.e., $\frac{R_r}{S} < 0$, exceeds the sum of the resistances of the armature and the network, there will be a negative resistance at the sub-synchronous frequency, and the sub-synchronous current will increase with time. This phenomenon, called in general Induction Generator Effect (IGE), only involves rotor electrical dynamics [35], and is referred to as SSIGE in this work, to emphasize the fact that it occurs at sub-synchronous (SS) frequency.

3.2.1 SSIGE MODES AND PARTICIPATION FACTORS

Participation factors provide a measure of the relative participation of each state variable to a given mode (eigenvalue) of the system. The magnitude of the normalized participation factor of the jth state variable in the ith mode, λ_i, of the system is defined as:

$$P_{ji} = \frac{|\Psi_{ji}||\Phi_{ij}|}{\sum\limits_{k=1}^{n} |\Psi_{jk}||\Phi_{kj}|} \tag{3.11}$$

where P_{ji} is the participation factor, n is the number of modes or state variables, and Ψ and Φ are right and left eigenvectors, respectively. For details, see [39].

Tables 3.1 and 3.2 show the eigenvalues and participation factors of the system when the wind speed is 7 m/s and the compensation level is 75%. The first table lists complex modes and the second table lists real modes. In these tables, the larger participation factors in each column are bolded, identifying the state variables that more strongly affect a given mode. By looking at these tables, one can readily find the participation of each state variable in a certain system mode. For example, based on Table 3.1 and using participation factors related to $\lambda_{9,10}$, one can see that this mode is associated primarily to three state variables, namely currents i_{qs}, i_{dr}, and DC link voltage, v_{DC}. Also, using Table 3.1 it can be observed that $\bar{\omega}_m$ and rotor-side converter PI-D have a high participation in mode $\lambda_{11,12}$. In Table 3.2, λ_{13} to λ_{22} are non-oscillatory and stable modes, and one can easily find the participation of each state variables on these modes by looking at this table. These modes will not be further discussed because they are not related to the SSR.

3.3 IDENTIFICATION OF SYSTEM MODES

In this section, the nature of modes $\lambda_{1,2}$, $\lambda_{3,4}$, $\lambda_{5,6}$, $\lambda_{7,8}$ is identified.

3.3.1 IDENTIFICATION OF SSR AND SUPSR MODES

Table 3.1 shows that modes $\lambda_{1,2}$ and $\lambda_{3,4}$ are primarily associated with state variables i_{qs}, i_{ds}, i_{qr}, and i_{dr}. Mode $\lambda_{1,2}$ has a frequency of 20.9947 Hz (or 131.913 rad./s) and can be identified to be the SSR mode (Mode 1). Similarly, mode $\lambda_{3,4}$ with a frequency of 98.23 Hz (or 617.197 rad./s) is the super-synchronous (SupSR) mode (Mode 2). This can be verified using Eq. 3.9, where f_n is calculated to be around 39 Hz. Given the synchronously rotating reference frame, the complementary SSR and SupSR frequencies are $f_s - f_n = 21$ Hz and $f_s + f_n = 99$ Hz, which matches the frequency of $\lambda_{1,2}$ and $\lambda_{3,4}$. Table 3.1 also shows that the SSR mode at 75%

Table 3.1: Participation factors of the system state variables for various system modes at 75% series compensation and 7 m/s wind speed (Part I, modes λ_1–λ_{12})

	$\lambda_{1,2}$ 0.128 ± j131.913	$\lambda_{3,4}$ -5.506 ± j617.197	$\lambda_{5,6}$ -9.911 ± j99.969	$\lambda_{7,8}$ -0.922 ± j5.999	$\lambda_{9,10}$ -875.428 ± j4217.869	$\lambda_{11,12}$ -0.115 ± j0.524
i_{qs}	**0.2791**	**0.2210**	**0.2783**	0.0031	**0.3435**	0.0102
i_{ds}	**0.2081**	**0.1755**	**0.2057**	0.0971	0.09770	0.0280
i_{0s}	0.0000	0.0000	0.0000	0.0000	0.0000	0.0000
i_{qr}	**0.2790**	**0.2027**	**0.2907**	0.0034	**0.1319**	0.0115
i_{dr}	**0.2082**	**0.1610**	**0.2172**	**0.1117**	0.0830	0.0321
i_{0r}	0.0000	0.0000	0.0000	0.0000	0.0000	0.0000
v_{cq}	0.0069	0.0671	0.0016	0.0030	0.0003	0.0008
v_{cd}	0.0066	0.0689	0.0015	0.0000	0.0000	0.0002
i_{lq}	0.0050	0.0487	0.0014	0.0000	0.0717	0.0003
i_{ld}	0.0056	0.0544	0.0017	0.0056	0.0051	0.0016
$\bar{\omega}_r$	0.0007	0.0000	0.0015	**0.3097**	0.0000	0.0594
$\bar{\omega}_m$	0.0000	0.0000	0.0000	0.0668	0.0000	**0.3131**
T_{tg}	0.0000	0.0000	0.0000	**0.3849**	0.0000	0.0167
v_{DC}	0.0000	0.0002	0.0000	0.0000	**0.2613**	0.0000
RSC-PI-A	0.0000	0.0000	0.0000	0.0000	0.0000	0.0001
RSC-PI-B	0.0000	0.0000	0.0000	0.0002	0.0000	0.0783
RSC-PI-C	0.0000	0.0000	0.0000	0.0000	0.0000	0.0019
RSC-PI-D	0.0000	0.0000	0.0000	0.0145	0.0000	**0.4388**
GSC-PI-A	0.0000	0.0000	0.0000	0.0000	0.0000	0.0000
GSC-PI-B	0.0000	0.0000	0.0000	0.0000	0.0058	0.0000
GSC-PI-C	0.0000	0.0000	0.0000	0.0000	0.0000	0.0064
GSC-PI-D	0.0003	0.0000	0.0000	0.0000	0.0001	0.0001

Table 3.2: Participation factors of the system state variables for various system modes at 75% series compensation and 7 m/s wind speed (Part II, modes λ_{13}–λ_{22})

	λ_{13} -2151.368	λ_{14} -114.749	λ_{15} -97.842	λ_{16} -0.500	λ_{17} -0.0143	λ_{18} -0.002	λ_{19} -19.929	λ_{20} -20.800	λ_{21} -0.000	λ_{22} -0.000
i_{qs}	**0.2381**	0.0344	0.0005	0.0000	0.0000	0.0000	0.0000	0.0000	0.0000	0.0000
i_{ds}	**0.1482**	**0.2785**	0.0689	0.0000	0.0000	0.0000	0.0000	0.0000	0.0000	0.0000
i_{0s}	0.0000	0.0000	0.0000	0.0000	0.0000	0.0000	**1.0000**	0.0000	0.0000	0.0000
i_{qr}	**0.2223**	0.0334	0.0003	0.0000	0.0000	0.0000	0.0000	0.0000	0.0000	0.0000
i_{dr}	**0.2214**	**0.2866**	0.0714	0.0000	0.0000	0.0000	0.0000	0.0000	0.0000	0.0000
i_{0r}	0.0000	0.0000	0.0000	0.0000	0.0000	0.0000	0.0000	**1.0000**	0.0000	0.0000
v_{cq}	0.0001	0.0035	0.0010	0.0000	0.0000	0.0000	0.0000	0.0000	0.0000	0.0000
v_{cd}	0.0015	0.0013	0.0009	0.0000	0.0000	0.0000	0.0000	0.0000	0.0000	0.0000
i_{lq}	0.0002	0.0006	0.0009	0.0000	0.0000	0.0000	0.0000	0.0000	0.0000	0.0000
i_{ld}	0.0980	0.0094	0.0029	0.0000	0.0000	0.0000	0.0000	0.0000	0.0000	0.0000
$\bar{\omega}_r$	0.0000	0.0000	0.0000	0.0000	0.0001	0.0001	0.0000	0.0000	0.0128	0.0461
$\bar{\omega}_m$	0.0000	0.0000	0.0000	0.0000	0.0008	0.0005	0.0000	0.0000	0.0612	**0.2198**
T_{tg}	0.0000	0.0000	0.0000	0.0000	0.0000	0.0000	0.0000	0.0000	0.0000	0.0000
v_{DC}	0.0430	0.0082	0.0154	0.0000	0.0000	0.0000	0.0000	0.0000	0.0000	0.0000
RSC-PI-A	0.0000	0.0000	0.0000	0.0000	0.0003	0.0003	0.0000	0.0000	0.2264	**0.7340**
RSC-PI-B	0.0000	0.0000	0.0000	0.0000	0.0107	0.0047	0.0000	0.0000	**0.6949**	0.0000
RSC-PI-C	0.0000	0.0000	0.0000	0.0000	0.0193	**0.9804**	0.0000	0.0000	0.0025	0.0000
RSC-PI-D	0.0000	0.0001	0.0000	0.0000	0.0071	0.0006	0.0000	0.0000	0.0000	0.0000
GSC-PI-A	0.0000	0.0000	0.0000	**0.9999**	0.0000	0.0000	0.0000	0.0000	0.0000	0.0000
GSC-PI-B	0.0011	0.0525	**0.7099**	0.0001	0.0000	0.0000	0.0000	0.0000	0.0000	0.0000
GSC-PI-C	0.0000	0.0000	0.0000	0.0000	**0.9614**	0.0132	0.0000	0.0000	0.0019	0.0000
GSC-PI-D	0.0258	**0.2911**	**0.1275**	0.0000	0.0000	0.0000	0.0000	0.0000	0.0000	0.0000

compensation and 7 m/s wind speed is unstable as the real part of this mode is positive, while the SupSR mode is stable.

3.3.2   IDENTIFICATION OF ELECTROMECHANICAL MODE

In order to identify the nature of mode $\lambda_{5,6}$, Table 3.3 shows this mode for different wind speeds and series compensation levels. The first row of this table shows the optimal shaft turbine speed in p.u. and the corresponding optimal frequency in Hz for each wind speed obtained using the MPPT plot shown in Figure 2.4. It is seen that the frequency of this mode varies with wind speed, while changing the compensation level has only a slight impact on the mode frequency.

Table 3.3: $\lambda_{5,6}$ at different wind speeds and compensation levels

	7 m/s (0.75 p.u./45 Hz)	8 m/s (0.85 p.u./51 Hz)	9 m/s (0.95 p.u./57 Hz)
75%	$-9.911 \pm$ j99.969	$-4.909 \pm$ j62.445	$-1.889 \pm$ j28.791
80%	$-12.767 \pm$ j99.942	$-5.498 \pm$ j62.995	$-2.123 \pm$ j29.335
90%	$-18.475 \pm$ j95.501	$-7.330 \pm$ j64.531	$-2.704 \pm$ j30.553

It can be observed that the frequency of this mode is the complementary of the frequency of the shaft turbine speed. For example, for wind speed equal to 7 m/s and compensation level equal to 75 %, the frequency of this mode is 99.97 $rad./s$ or 15.9 Hz, and its complementary is calculated to be 44.1 Hz (60 − 15.9 = 44.1 Hz). This frequency coincides with the frequency of the shaft turbine, i.e., 45 Hz. This same relationship applies to other wind speeds as well. Therefore, we can conclude that this mode is related to wind speed change and to mechanical dynamics. Also, using Table 3.1, it is observed that mode $\lambda_{5,6}$ is mostly associated with stator currents i_{qs} and i_{ds}, and to rotor currents i_{qr}, and i_{dr}. Therefore, this mode is related to both mechanical and electrical dynamics and is called electromechanical mode (Mode 3).

3.3.3   IDENTIFICATION OF SHAFT MODE

From Table 3.1, it is observed that the generator rotor speed $\bar{\omega}_r$ and the mechanical torque between two masses, T_{tg}, have the highest participation in mode $\lambda_{7,8}$. Therefore, mode $\lambda_{7,8}$ is related to the shaft mode (Mode 4). The shaft mode has low-frequency, about 0.954 Hz (or 5.999 rad./s). This mode at the present operating condition is stable, but it might become unstable if the series compensation level becomes too high, which will cause SSTI.

3.4   DISCUSSION OF THE SSIGE MODE

Table 3.4 shows the eigenvalues of the sub-synchronous resonance (SSR) and super - synchronous resonance (SupSR) modes of the system shown in Figure 2.1 for different series compensation levels and wind speeds. This table shows how the SSR and SupSR modes vary as a function of these two variables: (1) the wind speed V_ω and (2) the compensation level K. On the one hand, at a constant wind speed, when the compensation level increases, the stability of the SSR mode decreases while the stability of the SupSR mode slightly increases. Table 3.4 shows that the SSR mode is unstable for V_ω = 7 m/s and K = 55%, K = 60%, and K = 65%.

On the other hand, at a constant series compensation level, when the wind speed increases, the stability of both the SSR and SupSR modes increases. Based on Table 3.4, for K = 65% and V_ω = 7 m/s the SSR mode is highly unstable, but when V_ω increases, while K is kept constant, the stability of the SSR mode increases. For example, for K = 65% and V_ω = 8 m/s and 9 m/s, the SSR mode is stable.

Table 3.4: The SSR and SupSR modes of the system at different wind speeds V_ω and compensation levels K

V_ω (m/s)-K (%)	SSR Mode	SupSR Mode
7 - 50	-1.8784 ± j140.7799	-5.1561 ± j608.9960
7 - 55	**+1.2126 ± j128.5545**	-5.2253 ± j620.39633
7 - 60	**+5.9289 ± j118.8507**	-5.2812 ± j631.2477
7 - 65	**+9.6991 ± j112.3237**	-5.3158 ± j641.5941
8 - 55	-3.7739 ± j128.5441	-5.9986 ± j622.5840
8 - 60	-2.3818 ± j116.5455	-6.1252 ± j633.4910
8 - 65	-0.4696 ± j104.8237	-6.1877 ± j643.7831
9 - 55	-6.8362 ± j122.7589	-6.8150 ± j623.2366
9 - 60	-5.5889 ± 115.9793	-7.0351 ± j637.6388
9 - 65	-3.7165 ± j105.3277	-7.1718 ± j646.5196

Table 3.5: Rotor resistance under SSR and SupSR frequencies when the wind speed is kept constant at $V_\omega = 7\ m/s$ (45 Hz) and the compensation level changes

K (%)	$f_n\ Hz$	$\dfrac{R_r}{S_1}$	$\dfrac{R_r}{S_2}$
50%	37.59	-0.0278	0.00249
55%	39.54	-0.0397	0.00256
60 %	41.08	-0.0576	0.00262
65%	42.12	-0.0803	0.00265

3.4.1   EFFECT OF COMPENSATION LEVEL VARIATIONS

As mentioned in Section 3.4 above, the stability of SSIGE depends on both wind speed and compensation level. This section describes why increasing the compensation level decreases the stability of the SSR mode. In order to explain this fact, a specific example - where wind speed is kept constant at $V_\omega = 7\ m/s$, while the compensation level changes - is used. Using the MPPT curve of Figure 2.4, the electrical frequency corresponding to $V_\omega = 7\ m/s$ is found to be 45 Hz. Note that the value of rotor resistance of the DFIG used in this work is $R_r = 0.00549\ p.u.$, as listed in the Appendix A.

Table 3.5 shows the rotor resistances at sub-synchronous and and at super-synchronous frequencies for the aforementioned case. Note that if Table 3.4 is used to calculate f_n, since the models are built in a $d - q$ synchronous reference frame, the computed frequencies of the SSR and SupSR modes, given in Table 3.4, are $f_s - f_n$ and $f_s + f_n$, respectively. From Table 3.5, it can be easily observed that by increasing the compensation level, larger negative resistances are provided to the network, which decreases the stability of the SSR mode. The reason is that at a constant wind speed, or constant f_m, increasing the compensation level increases the electrical natural frequency of the system. Therefore, the absolute value of the DFIG slip S_1 under SSR frequency given in Eq. 3.10 decreases, providing more negative rotor resistance $\dfrac{R_r}{S_1}$ to the system. This decreases the stability of the SSR mode.

Table 3.6: Rotor resistance under SSR and SupSR frequencies when compensation level is kept constant at $K = 65\%$ ($f_n = 42.12\ Hz$) and wind speed changes

$V_\omega\ (m/s)$	$f_m\ Hz$	$\dfrac{R_r}{S_1}$	$\dfrac{R_r}{S_2}$
7	45	–0.0803	0.0026
8	51	–0.0260	0.0024
9	57	–0.0155	0.0023

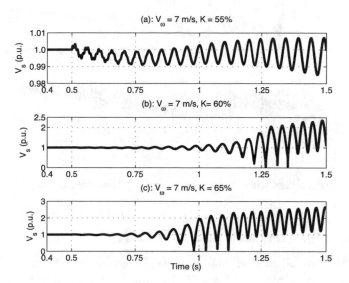

Figure 3.3: Terminal voltage when $V_\omega = 7\ m/s$ and (a) $K = 55\%$ (b) $K = 60\%$ (c) $K = 65\%$.

3.4.2   EFFECT OF WIND SPEED VARIATIONS

In order to explain the impact of wind speed variations on the stability of the SSR and SupSR modes, a specific example – where the compensation level is kept constant at $K = 65\%$ while the wind speed changes – is used. Using the MPPT curve shown in Figure 2.4, the electrical frequencies corresponding to different wind speeds are obtained. Table 3.6 shows the rotor resistances at sub-synchronous and at super-synchronous frequencies for this example. As it can be observed from this table, by increasing the wind speed, the SSR mode becomes more stable. The reason is that by increasing the wind speed – which, in other words, is equivalent to increasing the electrical frequency corresponding to wind speed f_m – the absolute value of the DFIG slip S_1 increases, providing less negative rotor resistance $\dfrac{R_r}{S_1}$ to the system. This increases the stability of the SSR mode.

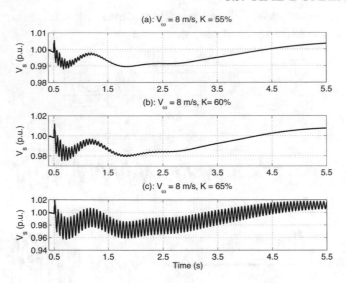

Figure 3.4: Terminal voltage when $V_\omega = 8 \ m/s$ and (a) $K = 55\%$ (b) $K = 60\%$ (c) $K = 65\%$.

3.5   TIME-DOMAIN SIMULATION

In order to confirm the eigenvalue analysis provided in Table 3.4, time domain simulations in PSCAD/EMTDC are performed. Figures 3.3 through 3.5 show the IG terminal voltage V_s for different wind speeds and compensation levels. Note that in the given simulation results, the system is first started with a lower series compensation level at which the wind farm is stable, i.e., $K = 50\%$, and then at $t = 0.5 \ s$, the compensation level is increased. The following conclusions can be drawn from the simulation results:

1. At lower wind speed, e.g., $V_\omega = 7 \ m/s$, when K increases, the stability of the SSR mode decreases, as seen in Figure 3.3.

2. The frequency of the oscillations in the time-domain simulation of Figures 3.3 - (a) - through - (c) is measured to be about 20 Hz, 18.18 Hz, and 17.85 Hz for $K = 55\%$, $K = 60\%$ and $K = 65\%$, respectively. These frequencies match the frequencies obtained using eigenvalue analysis for these cases given in Table 3.4. Note that both the eigenvalue analysis and the time-domain simulations are performed in a d-q synchronous reference frame.

3. From Figures 3.4 and 3.5, it is observed that increasing the wind speed stabilizes the SSR mode, as expected from Table 3.6. Additionally, these figures show that at a constant wind speed, increasing the compensation level, decreases the stability of the SSIGE mode, as discussed in reference to Table 3.5.

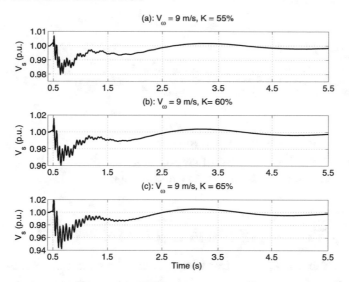

Figure 3.5: Terminal voltage when $V_\omega = 9\ m/s$ and (a) $K = 55\%$ (b) $K = 60\%$ (c) $K = 65\%$.

SUMMARY

In this chapter, the SSR phenomenon called induction generator effect (SSIGE) is studied, and the impact of wind farm parameters on this type of SSR is investigated using eigenvalue analysis and time-domain simulations in PSCAD/EMTDC. Based on the discussions given in this chapter, the following conclusions can be drawn:

1. The equivalent rotor resistance at sub-synchronous frequency has a negative value.

2. The SSIGE occurs when the absolute value of the equivalent rotor resistance at sub-synchronous frequency exceeds the sum of the positive resistances of the armature and the network.

3. At lower wind speeds and higher compensation levels, the possibility of the SSIGE in DFIG becomes higher.

4. The SSIGE is not related to the mechanical part of the system and is a purely electrical phenomenon.

CHAPTER 4

Torsional Interactions

This chapter focuses on the sub-synchronous torsional interactions (SSTI). First a descriptive definition is given; then, the real world possibility of the SSTI in DFIG wind farm is studied; finally, the impact of the stiffness coefficient and compensation level variations on this type of SSR is investigated.

4.1 WIND TURBINE DRIVE-TRAIN MODEL

In order to analyze the SSTI, it is better first to define the torsional frequencies of a DFIG wind turbine drive-train model. A common way is to represent the rotor as a number of discrete masses connected together by springs characterized by their damping and stiffness coefficients. Figure 4.1 shows the structure of a typical WTGS drive-train model. The equation of the i_{th} mass motion can be expressed as [61]:

$$2H_i \frac{d\Delta\omega_i}{dt} = T_i + T_{i,i+1} - T_{i,i-1} - D_i \frac{d\delta_i}{dt} \tag{4.1}$$

where

$$T_{i,j} = K_{i,j} \cdot (\delta_j - \delta_i) \tag{4.2}$$

$$\frac{d\delta_i}{dt} = \omega_i - \omega_r = \Delta\omega_i \tag{4.3}$$

If N discrete masses are considered, using Eqs. 4.1 through 4.3, a set of $2N$ differential equations can be obtained, which in a state-space description takes the following form [61]:

$$\dot{X} = AX + BU \tag{4.4}$$

where X and U are the state variables vector and the input torque vector, respectively.

The modes of the drive-train system, which are the eigenvalues of the state matrix A, are a set of complex-conjugate pairs of the form [61]:

$$\lambda_{i,i+1} = -\zeta_i \omega_{ni} \pm j\omega_{ni} \sqrt{1 - \zeta_i^2} \tag{4.5}$$

where ζ_i and ω_{ni} are the damping ratio and undamped natural frequency of the i_{th} mass.

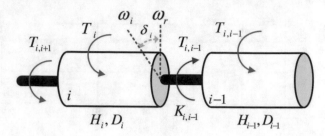

Figure 4.1: Structure of a typical drive-train model. $T_{i,i+1}$ = The torque applied to the i_{th} mass from $(i+1)_{th}$ mass, T_i = external torque applied to i_{th} mass, δ_i = torsional angle of the i_{th} mass, H_i inertia constant of the i_{th} mass, D_i = damping coefficient of the i_{th} mass, $K_{i,i-1}$ = stiffness coefficient between i_{th} and $(i-1)_{th}$ masses.

As a general case, a rotor with N masses has N modes, where $N-1$ modes represent the torsional modes of oscillation, and one remaining mode represents the oscillation of the entire rotor against the power system.

Using Eq. 4.5, the torsional natural frequency of the i_{th} mass can be obtained as [61]:

$$f_{mi} = \frac{\omega_{ni}\sqrt{1-\zeta_i^2}}{2\pi} \tag{4.6}$$

If the rotor of the generator oscillates at a torsional natural frequency, f_{mi}, this phenomenon induces an armature voltage component in the generator at two frequencies given by [62]:

$$f_{em_i} = f_s \pm f_{mi} \tag{4.7}$$

If f_{em_i} is close to f_n, which is the electrical natural frequency due to series compensation given by Eq. 3.9, the sub-synchronous torques generated by this sub-synchronous induced armature voltage can cause sustained mechanical oscillations of the rotor. This energy exchange between the electric part of the DFIG wind farm and its mechanical part is called *Torsional Interaction*, and it is referred to as SSTI in this work.

4.2   DOES SSTI OCCUR IN WIND FARMS?

In this section, we answer the question: "Does SSTI occur in wind farms?." The frequency of shaft torsional modes is a strong function of the shaft stiffness coefficient, i.e., $K_{i,j}$ in Figure 4.1. The values of $K_{i,j}$ in wind turbines are much smaller compared to the values found in steam, hydro, and diesel turbines. The typical value of $K_{i,j}$ reported in the literature is much less than 10 *p.u. Torque/rad* [61], while the values of $K_{i,j}$ for the different sections of a typical steam turbine reported in [35] are in the range 19 - 70 *p.u. Torque/rad*.

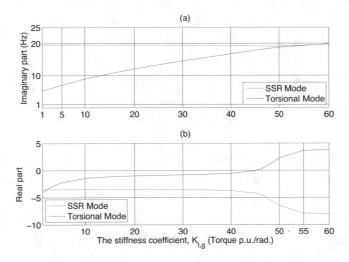

Figure 4.2: SSR and torsional modes versus the stiffness coefficient, $K_{t,g}$, when $V_\omega = 9\ m/s$ and $K = 55\%$: (a) Imaginary part (Hz) (b) Real part.

The low shaft stiffness coefficient in wind turbine drive-train leads to low torsional natural frequencies, which are in the range of 1–5 Hz. Therefore, based on the definition given for SSTI, in order to cause the SSTI in a wind farm, the electric natural frequency of the network should be in the range of 55–59 Hz. In order to obtain such a large electric natural frequency in the network, a very high series compensation level is needed, while in practice, the series compensation is normally not larger than 70%–75% for reasons such as load balancing with parallel paths, high fault current, and the possible difficulties of power flow control [12]. Hence, the SSTI is not a concern in WTGS. However, for the sake of completeness of the current work, the impact of shaft stiffness coefficients $K_{i,j}$ and series compensation level K on the SSTI mode is studied.

4.3   EFFECT OF SHAFT STIFFNESS COEFFICIENTS VARIATIONS

The studied WTGS shown in Figure 2.1 is composed of two masses, the generator and the turbine, and the stiffness coefficient between the turbine and the generator is called $K_{t,g}$. Also, the value of the stiffness coefficient in the studied system in this paper is $K_{t,g} = 0.15\ p.u.\ Torque/rad$. With this $K_{t,g}$, for $V_\omega = 9\ m/s$ and $K = 55\%$, the shaft torsional mode is calculated as $\lambda_{Torsional} = -3.2396 \pm j4.6767$, and this mode is stable. Using $\lambda_{Torsional}$, the torsional natural frequency is calculated to be less than 1 Hz. Therefore, in order to cause the SSTI in the system with the current $K_{t,g}$, f_n has to be about 59 Hz, which requires an unrealistically large compensation level.

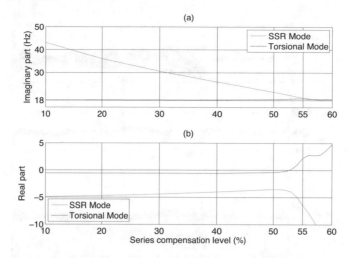

Figure 4.3: SSR and torsional modes versus series compensation level, K, when $V_\omega = 9\,m/s$: (a) Imaginary part (Hz) (b) Real part.

Figure 4.2 shows the SSR and torsional modes as a function of the stiffness coefficient, $K_{t,g}$, when $V_\omega = 9\,m/s$ and $K = 55\%$. As seen in this figure, by increasing $K_{t,g}$, as soon as the frequency of the torsional mode becomes close to the frequency of the SSR mode, the torsional mode becomes unstable. This occurs for a stiffness coefficient $K_{tg} = 45\,p.u.\,Torque/rad$, much larger than values typically encountered in wind turbines.

4.4   THE EFFECT OF SERIES COMPENSATION LEVEL VARIATIONS

In order to cause the SSTI in the wind farm, the value of $K_{t,g}$ is increased from 0.15 to 50 $p.u.\,Torque/rad$. Figure 4.3 shows the SSR and torsional modes as a function of compensation level when $V_\omega = 9\,m/s$. As seen in Figure 4.3 as long as the torsional natural frequency is not close to the SSR mode, the shaft mode is stable. Once the frequency of these modes become close to each other, the shaft mode becomes unstable

4.5   TIME-DOMAIN SIMULATION OF SSTI

In order to show the SSTI in the WTGS, time-domain simulations in PSCAD/EMTDC are performed. Figure 4.4 shows the system response including the torsional torque between masses I and II $T_{t,g}$, wind turbine speed ω_t, the electric torque T_e, and IG terminal voltage V_s, when $V_\omega = 9\,m/s$ and K changes. Note that in the given simulation results, the system is first started with a lower series compensation level, $K = 20\%$, and then at $t = 3\,s$ and $t = 8\,s$, the compen-

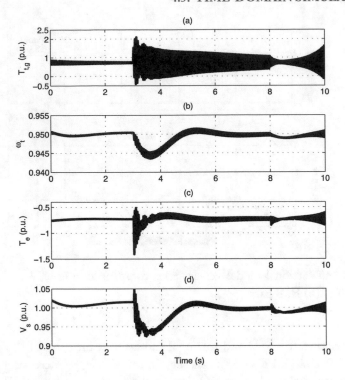

Figure 4.4: Time-domain simulations when $V_\omega = 9\ m/s$ and compensation level changes at times $t = 3\ s$ and $t = 8\ s$: (a) T_{tg} $(p.u.)$ (b) ω_t $(p.u.)$ (c) T_e $(p.u.)$ (d) V_s $(p.u.)$.

sation level is increased to $K = 50\%$ and $K = 55\%$, respectively. In these simulations, $K_{t,g} = 50$ $p.u.\ Torque/rad$. The following conclusions can be drawn from the simulation results:

1. The wind farm is stable at lower compensation levels, i.e., when $K = 20\%$ and $K = 50\%$, as expected from Figure 4.3. However, when the compensation level increases to 55%, the SSTI occurs in the WTGS, and the wind farm goes unstable due to the unstable torsional mode.

2. Even when the torsional mode is stable at lower compensation levels, i.e., when $K = 20\%$ and $K = 50\%$, the SST-TI is very lightly damped. This is due to the fact that the damping ratio of the torsional mode is very small even at these compensation levels, i.e., 0.5% and 0.14% for $K = 20\%$ and $K = 50\%$, respectively.

3. Therefore, in some cases, depending on system parameters, the torsional interaction mode may have a low damping ratio and an SSR damping controller may be desirable.

SUMMARY

In this chapter, the sub-synchronous torsional interactions (SSTI) case is studied, and the impact of wind farms parameters on this type of SSR is investigated using eigenvalue analysis and time domain simulations in PSCAD/EMTDC. Based on the discussions given in this chapter, the following conclusions can be drawn:

1. The SSTI may occur if the complement of one of the torsional natural frequencies of the drive-train turbine system is close to the electric natural frequency.

2. Because of the low values of shaft stiffness coefficient in WTGS, the SSTI is not a concern in practice.

CHAPTER 5

Control Interactions

5.1 CONTROL INTERACTIONS (SSCI)

Sub-synchronous control interactions (SSCI) are mainly due to the interactions between DFIG wind turbine controllers and the series compensated transmission line, to which the wind farm is connected. Unlike the aforementioned SSR types, the SSCI does not have well-defined frequencies of concern due to the fact that the frequency of oscillations in SSCI depends not only on the configuration of the series compensated transmission line and induction generator parameters, but also on the wind turbine controller configuration and parameters [27] - [31]. Moreover, the oscillations caused by the SSCI may grow faster compared to previously mentioned SSR types, since the undamped oscillation in SSCI completely depends on the electrical and controller interactions, which have a smaller time constant.

The SSCI has come into prominence since the ERCOT event of 2009 [27] - [31]. A faulted line and subsequent outage in the network caused a large DFIG wind farm to become radially connected to the series compensation network, resulting in rapidly increasing sub-synchronous frequency oscillations leading to damage to both the series capacitor and the wind turbine [27] - [31]. The SSCI system can be simplified as shown in Figure 5.1. According to this figure, as mentioned earlier, the reason for the SSCI is the interaction between the DFIG controllers and the network electric natural frequency.

5.2 EXISTING AND PLANNED SERIES COMPENSATED WIND FARMS

In 2005 [63], [64], the public utility commission of Texas (PUCT) developed a plan to build 2300 miles of new 345 kV transmission lines to accommodate an increase of 11553 MW of wind energy in West Texas. Some of the transmission lines in the plan were designed to have 50% series compensation. In Figure 5.2, a section of the Electric Reliability Council of Texas (ERCOT) grid is shown, where a 200 MW DFIG wind farm is connected to Bus 2. The nominal voltage in all buses is 138 kV, except for buses 12-16, where transformers step up the voltage level from 138 kV to 345 kV. The series compensation capacitors are located on the Bus 13-Bus 16 transmission line and on the Bus 15-Bus 16 transmission line, with compensation levels from 50% to 80%. The thick green line in Figure 5.2 is the worst case scenario in terms of susceptibility to SSR, where all other lines in the network are open, and thereby, the wind farm is radially connected to the series compensated lines via Bus 2, Bus 3, Bus 8, Bus 13, Bus 16 and Bus 15. In the ERCOT

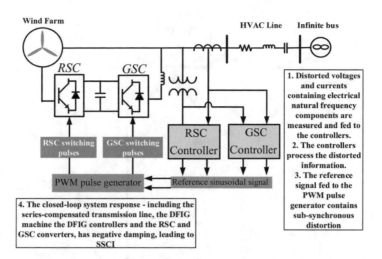

Figure 5.1: The closed-loop system response - including the series-compensated transmission line, the DFIG machine the DFIG controllers and the RSC and GSC converters, has negative damping, leading to SSCI.

event of 2009 [27] - [31], a faulted line and subsequent outage in the network caused the DFIG wind farm to become radially connected to the series compensation network. In this case, the power network shown in Figure 5.2 is reduced to a radial single-machine-infinite bus series-compensated network. This event resulted in rapidly increasing of sub-synchronous frequency oscillations leading to damage to both the series capacitor and the wind turbine [27] - [31]. Note that this case of a wind farm radially connected to series-compensated transmission lines is similar to the system studied in this work shown in Figure 2.1. This shows the practical relevance of the research presented in this work.

Additionally, with rapid increase of wind power energy in Southern Minnesota and South Dakota, the Xcel Energy Inc. has planned (or already implemented) series compensation in several transmission lines, including a 150-MW DFIG wind farm connected to a 54-mile-long, 60% series compensated, 345 kV Wilmarth (WLM) - Lakefield Generating station (LFD) transmission line, as seen in Figure 5.3 [34], [65]. A switching event around the series compensated transmission line connected to wind farm and combustion turbine generation resulted in growing unstable sub-synchronous oscillations [34].

Moreover, in [66], ABB Inc. has performed a "Dakotas Wind Transmission Study" to investigate the transmission line capacity for up to 500 MW of new wind generation planned to be located at seven different sites. The results revealed that the peak wind generated power cannot be delivered because the uncompensated transmission line exhibited congestion. The report suggests that providing 35% and 50% series compensation for the existing transmission lines can eliminate

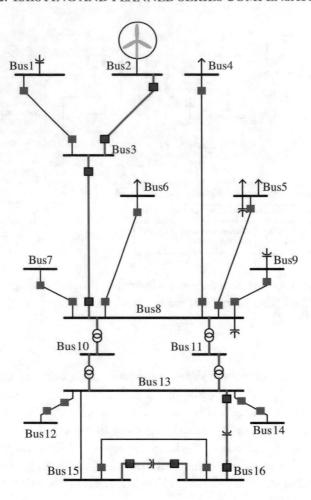

Figure 5.2: Single line diagram of a part of ERCOT grid, where a 200 MW DFIG wind farm is connected to the Bus 2 [27],[30].

power congestion and allow the wind-generated power to be exploited. The ABB Inc. report also states that special studies must be performed in order to avoid the SSR in the system.

In addition, [67] discusses technical requirements for the interconnection to Bonneville Power Administration (BPA), in Pacific Northwest, transmission grid, including series compensation to transmit wind power energy. Also, [68] gives a report regarding reinforcement of transmission lines of Alberta Electric System Operator (AESO) using series compensation. Some of these lines are connected directly or indirectly to wind farms.

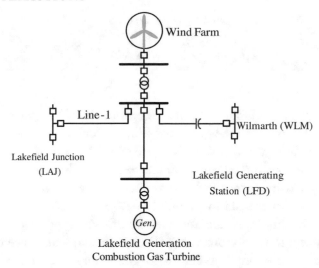

Figure 5.3:   Single line diagram of the 54 mile 345 KV Wilmarth (WLM)- Lakefield Generating station (LFD) transmission line connected to the wind farm [34], [65].

SUMMARY

In this chapter, sub-synchronous control interactions (SSCI) are briefly discussed. A more detailed study of the SSCI is left as future work. Based on the discussion given in this chapter, the following conclusions can be drawn:

1. The SSCI is an interaction between the DFIG wind turbine controllers and the series compensated transmission line, to which the wind farm is radially connected.

2. The SSCI does not have well-defined frequencies of concern.

3. The oscillations caused by the SSCI may grow faster compared to SSIGE and SSTI.

Conclusion

Worldwide rapid penetration of wind power into electric power grids makes it necessary for the power utilities to transmit the generated wind power without congestion. In addition, the global trend toward a deregulated power market requires that any solution to increase the power transfer capability of an existing transmission line be financially competitive. In many cases, cost effective series compensation increases the transmissible power of an existing transmission line at a fraction of the cost and required time to build a new transmission line.

However, a factor hindering the extensive use of series capacitive compensation is the potential risk of sub-synchronous resonance (SSR), which may cause severe damage in the wind farm, if not prevented. For example, in 2009 a SSR event happened in Electric Reliability Council of Texas (ERCOT) electric grid, leading to damage to both the series capacitor and the wind turbines. Therefore, SSR studies must be performed before the implementation of series compensation in a transmission line. These studies include small-signal analysis of the system, to which the wind farm is connected, in order to identify the eigenvalues of the system, especially the SSR mode.

In this regard, this book has developed a comprehensive approach to small-signal stability analysis of a doubly-fed induction generator (DFIG)-based series compensated wind farm using Matlab/Simulink. The developed approach can be easily extended to the small signal stability analysis of more complex power systems. The developed model can be used to obtain eigenvalues of the system and to help design stabilizing controllers using the eigenvalue analysis method. Moreover, the various SSR types that may occur in wind farm have been briefly reviewed, and the impact of the different parameters of the system, such as wind speed and compensation level, on the various SSR types has been investigated.

List of Parameters of the Power System Under Study

Table A.1: Parameters of the single 2 MW and 100 MW aggregated DFIG. Values are in (p.u.), unless otherwise indicated.

Base Power	2 MW	100 MW
Based voltage (V_{LL})	690 V	690 V
X_{ls}	0.09231	0.09231
X_{lr}	0.09955	0.09955
:w R_s	0.00488	0.00488
R_r	0.00549	0.00549
X_{tg}	0.3 (0.189 mH)	0.3 (0.189/50 mH)
DC-link base base voltage	1200 V	1200 V
DC-link capacitor	14000 μF	50*14000 μF

Table A.2: Parameters of the network and shaft system. Values are in ($p.u.$).

R_L	0.02	X_L	0.50	X_T	0.14	X_{sys}	0.06
H_t	4.29 s	H_g	0.9 s	D_{tg}	1.5	K_{tg}	0.15

Bibliography

[1] M. Islam, H. A. Mohammadpour, A. Ghaderi, C. Brice, Y. J. Shin, "Time-frequency based instantaneous power components for transient disturbances according to IEEE standard 1459," *IEEE Transactions on Power Delivery*, doi: 10.1109/TPWRD.2014.2361203. DOI: 10.1109/TPWRD.2014.2361203.

[2] M. Islam, H. A. Mohammadpour, P. Stone, Y. J. Shin, "Time-frequency based power quality analysis of variable speed wind turbine generators," *IEEE 39th Annual Conference of the Industrial Electronics Society, (IECON) 2013*, pp. 6426–6431, 10–13 Nov., Vienna, Austria. DOI: 10.1109/IECON.2013.6700194.

[3] H. A. Mohammadpour, E. Santi, "SSR Damping Controller Design and Optimal Placement in Rotor-Side and Grid-Side Converters of Series Compensated DFIG-Based Wind Farm," *IEEE Transactions on Sustainable Energy*, doi: 10.1109/TSTE.2014.2380782. DOI: 10.1109/TSTE.2014.2380782. 5, 6, 9, 10, 11

[4] H. A. Mohammadpour, A. Ghaderi, E. Santi, "Analysis of sub-synchronous resonance in doubly-fed induction generator-based wind farms interfaced with gate-controlled series capacitor," *IET Generation, Transmission & Distribution*, vol. 8, no. 12, pp.1998–2011, December 2014. DOI: 10.1049/iet-gtd.2013.0643. 16

[5] T. Ackermann, *Wind power in power systems*, England, John Wiley & Sons, 2005. DOI: 10.1109/TIE.2010.2053340.

[6] N. Denniston, A. M. Massoud, S. Ahmed, P. N. Enjeti,"Multiple-module high-gain high-voltage DC-DC transformers for offshore wind energy systems," *IEEE Trans. Industrial Electronics*, vol. 58, no. 5, pp.1877–1886, May 2011. DOI: 10.1109/TIE.2010.2053340.

[7] R. Teixeira Pinto, P. Bauer, S. F. Rodrigues, E. J. Wiggelinkhuizen, J. Pierik, B. Ferreira, "A novel distributed direct-voltage control strategy for grid integration of offshore wind energy systems through MTDC network," *IEEE Trans. Industrial Electronics*, vol. 60, no. 6, pp. 2429–2441, June 2013. DOI: 10.1109/TIE.2012.2216239.

[8] Soledad Bernal-Perez, Salvador Ano-Villalba, Ramon Blasco-Gimenez, "Efficiency and fault ride-through performance of a diode-rectifier- and VSC-inverter-based HVDC link for offshore wind farms," *IEEE Trans. Industrial Electronics*, vol. 60, no. 6, pp. 2401–2409, June 2013. DOI: 10.1109/TIE.2012.2222855.

[9] R. Blasco, N. Aparicio, S. Ano-Villalba, S. Bernal-Perez, "LCC-HVDC connection of off-shore wind farms with reduced filter banks," *IEEE Trans. Industrial Electronics*, vol. 60, no. 6, pp. 2372–2380, June 2013. DOI: 10.1109/TIE.2012.2227906.

[10] C. J. Chou, Y. K. Wu, Member, G. Y. Han, C. Y. Lee, "Comparative evaluation of the HVDC and HVAC links integrated in a large offshore wind farm: an actual case study in Taiwan," *IEEE Transactions on Industry Applications*, vol. 48, no. 5, pp. 1639–1648, September 2012. DOI: 10.1109/TIA.2012.2209622.

[11] "Series compensation: boosting transmission capacity," http://www.abb.com/FACTS.

[12] N. G. Hingorani and L. Gyugi, *Understanding FACTS*. Piscataway, NJ, 656 USA: IEEE Press, 2000. 25, 26, 37

[13] Esmaeilian, A.; Ghaderi, A.; Tasdighi, M.; Rouhani, A.,"Evaluation and performance comparison of power swing detection algorithms in presence of series compensation on transmission lines," Environment and Electrical Engineering (EEEIC), 2011 10th International Conference on, vol. , no. , pp. 1,4, 8–11 May 2011. DOI: 10.1109/EEEIC.2011.5874850.

[14] H. A. Mohammadpour, M. R. Pahlavani, A. Shoulaie, "On control of gate controlled series capacitor for SSR and power oscillation damping," *Compatibility and Power Electronics, 2009. CPE'09. IEEE*, pp. 196–203, May 2009. DOI: 10.1109/CPE.2009.5156035.

[15] H. A. Mohammadpour, M. R. Pahlavani, A. Shoulaie, "On harmonic analysis of multi-module gate-controlled series capacitor (MGCSC) considering SSR phenomenon," *International Review of Electrical Engineering*, vol. 4, no. 4, pp. 627–634, Aug. 2009.

[16] M. Pahlavani, H. A. Mohammadpour, "Damping of sub-synchronous resonance and low-frequency power oscillation in a series-compensated transmission line using gate-controlled series capacitor," *Electric Power Systems Research*, vol. 81, no. 2, pp. 308–317, Feb. 2011. DOI: 10.1016/j.epsr.2010.09.007.

[17] H. A. Mohammadpour, Y. J. Shin, E. Santi, "SSR analysis of a DFIG-based wind farm interfaced with a gate-controlled series capacitor," *Applied Power Electronics Conference and Exposition (APEC), 2014 Twenty-Ninth Annual IEEE*, pp. 3110–3117, March 2014. DOI: 10.1109/APEC.2014.6803749.

[18] M. Mokhtari, J. Khazaie, D. nazarpour , "Sub-Synchronous Resonance damping via Doubly Fed Induction Generator," *International Journal of Electrical Power & Energy Systems*, vol. 53, no. 1, pp. 876–883, Dec. 2013. DOI: 10.1016/j.ijepes.2013.05.054.

[19] J. Khazaie, M. Mokhtari, M. Khalilyan, D. Nazarpour , "Sub-Synchronous Resonance damping using Distributed Static Series Compensator (DSSC) enhanced with fuzzy logic controller," *International Journal of Electrical Power & Energy Systems*, vol. 43, no. 1, pp. 80–89, Dec. 2012. DOI: 10.1016/j.ijepes.2012.05.009.

[20] H.A. Mohammadpour, S.M.H. Mirhoseini, A. Shoulaie, "Comparative study of proportional and TS fuzzy controlled GCSC for SSR mitigation," *Power Engineering, Energy and Electrical Drives, 2009. POWERENG '09. International Conference on*, pp. 564–569, March 2009. DOI: 10.1109/POWERENG.2009.4915145.

[21] IEEE SSR Working Group, "Terms, definitions and symbols for subsynchronous oscillations," *IEEE Trans. Power Appl. Syst.* vol. PAS-104, no. 6, pp. 1326–1334, June 1985. DOI: 10.1109/TPAS.1985.319152.

[22] H. A. Mohammadpour, E. Santi, "Modeling and control of gate - controlled series capacitor interfaced with a DFIG-based wind farm," *IEEE Transactions on Industrial Electronics*, DOI:10.1109 / TIE.2014.2347007, Available on-line: 12 August 2014. DOI: 10.1109/TIE.2014.2347007. 5, 6, 8, 9, 10, 11, 12, 14, 15

[23] S. O. Faried, I. Unal, D. Rai, J. Mahseredjian, "Utilizing DFIG-based wind farms for damping subsynchronous resonance in nearby turbine-generators," *IEEE Trans. Power Systems*, vol. 28, no. 1, pp. 452–459, Feb. 2013. DOI: 10.1109/TPWRS.2012.2196530.

[24] K. R. Padiar, *Analysis of Sub-synchronous Resonance in Power Systems*. Kluwer Academic Publishers (KAP), Boston, 1999. DOI: 10.1007/978-1-4615-5633-6.

[25] H. A. Mohammadpour, Y. J. Shin, E. Santi, "SSR analysis of a DFIG-based wind farm interfaced with a gate-controlled series capacitor," *IEEE Twenty-Ninth Annual Applied Power Electronics Conference and Exposition (APEC) 2014*, pp. 3110–3117, 16–20 March, Fort Worth, TX, USA. DOI: 10.1109/APEC.2014.6803749.

[26] H. A. Mohammadpour, Md. Islam, D. Coats Y. J. Shin, E. Santi, "Sub-synchronous resonance mitigation in wind farms using gate-controlled series capacitor," *Power Electronics for Distributed Generation Systems (PEDG), 2013 4th IEEE International Symposium on*, pp. 1–6, July 2013. DOI: 10.1109/PEDG.2013.6785619.

[27] M. Sahni, D. Muthumuni, B. Badrzadeh, A. Gole, A. Kulkarni, "Advanced screening techniques for sub-synchronous interaction in wind farms," *Transmission and Distribution Conference and Exposition (T&D)*, pp. 1–9, May 2012. DOI: 10.1109/TDC.2012.6281671. 41, 42, 43

[28] H. A. Mohammadpour, E. Santi, "Sub-synchronous resonance analysis in DFIG-based wind farms: definitions and problem identification - Part I," *IEEE Energy Conversion Congress and Exposition (ECCE) 2014*, pp. 1–8, 14–18 September, Pittsburgh, PA , USA. DOI: 10.1109/ECCE.2014.6953480.

[29] H. A. Mohammadpour, E. Santi, "Sub-synchronous resonance analysis in DFIG-based wind farms: mitigation methods - TCSC, GCSC, and DFIG controllers - Part II," *IEEE*

Energy Conversion Congress and Exposition (ECCE), pp. 1–8, 14–18 September, Pittsburgh, PA , USA. DOI: 10.1109/ECCE.2014.6953603.

[30] M Sahni, B Badrzadeh, D Muthumuni, Y.Cheng, H.Yin, S-H. Huang, Y. Zhou, "Sub-synchronous interaction in wind power plants- part II: an ERCOT case study," *Power and Energy Society General Meeting, 2012 IEEE*, pp. 1–9, July 2012. DOI: 10.1109/PESGM.2012.6345355. 43

[31] Garth D. Irwin, Amit K. Jindal, Andrew L. Isaacs, "Sub-synchronous control interactions between type 3 wind turbines and series compensated AC transmission systems," *Power and Energy Society General Meeting, 2011 IEEE*, pp. 1–6, July 2011 DOI: 10.1109/PES.2011.6039426. 41, 42

[32] Yunzhi Cheng, Mandhir Sahni, Dharshana Muthumuni, Babak Badrzadeh, "Reactance scan crossover-based approach for investigating SSCI concerns for DFIG-based wind turbines," *IEEE Transactions on Power Delivery*, vol. 28, no. 3, pp. 742–752, April 2013. DOI: 10.1109/TPWRD.2012.2223239.

[33] Liang Wang, Xiaorong Xie, Qirong Jiang, H.R. Pota, "Mitigation of multimodal subsynchronous resonance via controlled injection of super-synchronous and sub-synchronous currents," *Power Systems, IEEE Transactions on*, vol. 29, no. 3, pp. 1335,1344, May 2014. DOI: 10.1109/TPWRS.2013.2292597.

[34] A. E. Leon, J. A. Solsona,"Sub-synchronous interaction damping control for DFIG wind turbines," *IEEE Trans. Power Systems*, vol., no. 99, pp. 1–10, June 2014. DOI: 10.1109/TP-WRS.2014.2327197. 42, 44

[35] IEEE SSR Task Force, "First benchmark model for computer simulation of subsynchronous resonance," *IEEE Trans. Power App. Syst.*, vol. PAS-96, pp. 1562–1572, Sep./Oct. 1997. DOI: 10.1109/T-PAS.1977.32485. 5, 6, 28, 36

[36] L. Fan, Z. Miao, "Mitigating SSR using DFIG-based wind generation," *IEEE Trans. Sustainable Energy*, vol. 3, no. 3, pp. 349–358, July 2012. DOI: 10.1109/TSTE.2012.2185962. 5

[37] M. Jafar, M. Molinas, T. Isobe, R. Shimada, "Transformer-less series reactive/harmonic compensation of line-commutated HVDC for offshore wind power integration," *Power Delivery, IEEE Transactions on*, vol. 29, no. 1, pp. 353,361, Feb. 2014 DOI: 10.1109/TP-WRD.2013.2270090. 5

[38] M. J. Hossain, T. K. Saha, N. Mithulananthan, H. R. Pota, "Control strategies for augmenting LVRT capability of DFIGs in interconnected power systems," *IEEE Trans. Industrial Electronics*, vol. 60, no. 6, pp. 2510–2522, June 2013. DOI: 10.1109/TIE.2012.2228141. 5

[39] P. Kundur, *Power System Sability and Control.*, New York: McGraw Hill, 1994. 5, 6, 28

[40] P. C. Krause, O. Wasynczuk, S. D. Sudhoff, *Analysis of Electric Machinery,*, IEEE Press, Piscataway, NJ, 1995. 7, 11, 12, 15, 16

[41] L. Wang, M. Sa-Nguyen Thi, "Stability analysis of four PMSG-based offshore wind farms fed to an SG-based power system through an LCC-HVDC link," *IEEE Trans. Industrial Electronics*, vol. 60, no. 6, pp. 2392–2400, June 2013. DOI: 10.1109/TIE.2012.2227904. 8

[42] H. Akagi and H. Sato, "Control and performance of a doubly-fed induction machine intended for a flywheel energy storage system," *IEEE Trans. Power Electron.*, vol. 17, no. 1, pp. 109–116, Jan. 2002. DOI: 10.1109/63.988676. 9

[43] R. Pena, J. C. Clare, G. M. Asher, "A doubly fed induction generator using back-to-back PWM converters supplying an isolated load from a variable speed wind turbine," *IEE Proc. Electr. Power Appl.*, vol. 143, no. 5, pp. 380–387, 1996. DOI: 10.1049/ip-epa:19960454. 9

[44] M. Rahimi and M. Parniani, "Efficient control scheme of wind turbines with doubly fed induction generators for low voltage ride-through capability enhancement," *IET Renewable Power Gener.*, vol. 4, no. 3, pp. 242–252, May 2010. DOI: 10.1049/iet-rpg.2009.0072.

[45] P. Ledesma, J. Usaloa, "Effect of neglecting stator transients in doubly fed induction generators models," *IEEE Trans. Energy Convers.*, vol. 19, no. 2, pp. 459–461, June 2004. DOI: 10.1109/TEC.2004.827045. 9

[46] J. Arbi, M. J.-B. Ghorbal, I. Slama-Belkhodja, L. Charaabi, "Direct virtual torque control for doubly fed induction generator grid connection," *IEEE Trans. Ind. Electron.*, vol. 56, no. 10, pp. 4163–4173, Oct. 2009. DOI: 10.1109/TIE.2009.2021590. 9

[47] E. Tremblay, S. Atayde, A. Chandra, "Comparative study of control strategies for the doubly fed induction generator in wind energy conversion systems: A DSP-based implementation approach," *IEEE Trans. Sustain. Energy*, vol. 2, no. 3, pp. 288–299, July 2011. DOI: 10.1109/TSTE.2011.2113381.

[48] G. Abad, M. A. Rodriguez, J. Poza, "Two-level VSC based predictive direct torque control of the doubly fed induction machine with reduced torque and flux ripples at low constant switching frequency," *IEEE Trans. Power Electron.*, vol. 23, no. 3, pp. 1050–1060, May 2008. DOI: 10.1109/TPEL.2008.921160. 9

[49] G. Abad, M. A. Rodriguez, G. Iwanski, J. Poza, "Direct power control of doubly-fed-induction-generator-based wind turbines under unbalanced grid voltage," *IEEE Trans. Power Electron.*, vol. 25, no. 2, pp. 442–452, Feb. 2010. DOI: 10.1109/TPEL.2009.2027438. 9

[50] D. Santos-Martin, J. L. Rodriguez-Amenedo, S. Arnalte, "Direct power control applied to doubly fed induction generator under unbalanced grid voltage conditions," *IEEE Trans. Power Electron.*, vol. 23, no. 5, pp. 2328–2336, Sep. 2008. DOI: 10.1109/TPEL.2008.2001908.

[51] L. Xu, P. Cartwright, "Direct active and reactive power control of DFIG for wind energy generation," *IEEE Trans. Energy Convers.*, vol. 21, no. 3, pp. 750–758, Sep. 2006. DOI: 10.1109/TEC.2006.875472. 9

[52] M. Tazil, V. Kumar, R. C. Bansal, S. Kong, Z. Y. Dong, W. Freitas, "Three-phase doubly fed induction generators: An overview," *IET Elect. Power Appl.*, vol. 4, no. 2, pp. 75–89, Feb. 2010. DOI: 10.1049/iet-epa.2009.0071. 9

[53] R. Datta, V. T. Ranganathan, "Direct power control of grid-connected wound rotor induction machine without rotor position sensors," *IEEE Trans. Power Electron.*, vol. 16, no. 3, pp. 390–399, May 2001. DOI: 10.1109/63.923772. 9

[54] S. Tohidi, H. Oraee, M. R. Zolghadri, S. Shao, and P. Tavner, "Analysis and enhancement of low-voltage ride-through capability of brushless doubly fed induction generator," *IEEE Trans. Ind. Electron.*, vol. 60, no. 3, pp. 1146–1155, Mar. 2013. DOI: 10.1109/TIE.2012.2190955. 9, 10, 11

[55] F. Mei, B. C. Pal, " Modeling of doubly-fed induction generator for power system stability study," in *Proc. IEEE Power & Energy General Meeting 2008*, Pittsburgh, PA, Jul. 2008. DOI: 10.1109/PES.2008.4596214.

[56] F. Mei, B. C. Pal, "Modeling adequacy of the doubly fed induction generator for small-signal stability studies in power systems," *IET Renewable Power Generation*, vol. 2, no. 3, pp. 181–190, Sep. 2008. DOI: 10.1049/iet-rpg:20070128. 9

[57] O. A. Lara, N. Jenkins, J. Ekanayake, P. Cartwright, M. Hughes, *Wind Energy Generation Modeling and Control*, England, John Wiley & Sons, 2009. 9

[58] Y. Lei, A. Mullane, G. Lightbody, R. Yacamini, "Modeling of the wind turbine with a doubly fed induction generator for grid integration studies," *IEEE Trans. Energy Conversion*, vol. 21, no. 1, pp. 257–264, March 2006. DOI: 10.1109/TEC.2005.847958. 11, 14

[59] H. Klee, R. Allen, *Simulation of Dynamic Systems with MATLAB and SIMULINK, Taylor & Francis Group.*, second edition, Boca Raton, FL, 2011. 19

[60] P. M. Anderson, B. L. Agrawal, J. E. Van Less, *Sub-synchronous Resonance in Power Systems.* IEEE Press, New York, 1990. 26, 27

[61] Z. Lubonsy, *Wind Turbine Operation in Electric Power Systems.* Springer, New York, 2010. 35, 36

[62] Ulf Häger, Christian Rehtanz, *Monitoring, Control and Protection of Interconnected Power Systems*. Springer, 2014. 36

[63] "competitive renewable energy zones (CREZ) transmission optimization study," http://www.ercot.com/news/presentations/2008. 41

[64] "CREZ reactive power compensation study," http://new.abb.com/products/power-consulting/reference-crez. 41

[65] K. Narendra, D. Fedirchuk, R. Midence, N. Zhang, A. Mulawarman, P. Mysore, V. Sood, "New microprocessor based relay to monitor and protect power systems against subharmonics," *in Proc. IEEE Elect. Power and Energy Conf. (EPEC'11)*, 2011, pp. 438–443. 42, 44

[66] "Dakotas wind transmission study summary, Task 1 through Task 4", http://www.wapa.gov/ugp/PlanProject/. 42

[67] "Technical requirements for interconnection to the BPA transmission grid STD-N-000001" http://www.bpa.gov/transmission/Pages/default.aspx. 43

[68] "Southern Alberta Transmission Reinforcement," http://www.aeso.ca/transmission/16869.html. 43

Authors' Biographies

HOSSEIN ALI MOHAMMADPOUR

Hossein Ali Mohammadpour (IEEE S '2009– M '2015) received the B.Sc. and M.Sc. degrees all in electrical engineering power systems from the Iran University of Science and Technology (IUST), Tehran, Iran, in 2006 and 2009, respectively. He also received the Ph.D. degree in electrical engineering with focus on renewable energy systems from the University of South Carolina, Columbia, SC, USA, in December 2014, where he continued as a postdoctoral fellow until March 2015.Since March 2015, he has been with the NRG Renew, Scottsdale AZ, USA as a Senior Power Systems Engineer.

Dr. Mohammadpour has published over fifty papers in power systems, power electronics, and renewable energy resources in international journals and conference proceedings. His current research interests include power systems stability and control, micro-grid systems, photovoltaic, battery energy storage systems, and smart grid.

ENRICO SANTI

Enrico Santi received the Dr. Ing. degree in electrical engineering from the University of Padua, Italy in 1988 and the M.S. and Ph.D. degrees from Caltech in 1989 and 1994, respectively. He worked as a senior design engineer at TESLAco from 1993 to 1998, where he was responsible for the development of various switching power supplies for commercial applications. Since 1998 he has been with the University of South Carolina where he is currently an associate professor in the electrical engineering department.

Dr. Santi has published over one hundred and fifty papers in power electronics and modeling and simulation in international journals and conference proceedings, is co-author of two books and holds three patents. His research interests include switched-mode power converters, advanced modeling and simulation of power systems, modeling and simulation of semiconductor power devices, control of power electronics systems.